Soft, Biological and Composite Nanomaterials

Soft, Biological and Composite Nanomaterials

Editors

Beom Soo Kim
Arvind Gupta

MDPI • Basel • Beijing • Wuhan • Barcelona • Belgrade • Manchester • Tokyo • Cluj • Tianjin

Editors
Beom Soo Kim
Chungbuk National University
Republic of Korea

Arvind Gupta
Chungbuk National University
Republic of Korea

Editorial Office
MDPI
St. Alban-Anlage 66
4052 Basel, Switzerland

This is a reprint of articles from the Special Issue published online in the open access journal *Nanomaterials* (ISSN 2079-4991) (available at: https://www.mdpi.com/journal/nanomaterials/special_issues/soft_biological_composite).

For citation purposes, cite each article independently as indicated on the article page online and as indicated below:

LastName, A.A.; LastName, B.B.; LastName, C.C. Article Title. *Journal Name* **Year**, *Article Number*, Page Range.

ISBN 978-3-03943-092-5 (Hbk)
ISBN 978-3-03943-093-2 (PDF)

Contents

About the Editors

Beom Soo Kim is a Professor of Chemical Engineering at Chungbuk National University, Cheongju, Korea. He studied chemical engineering at Seoul National University (1988), obtained a PhD in biochemical engineering at KAIST (1993), and completed postdoctoral work at MIT (1998). He started his lab at Chungbuk National University (2001) and spent sabbatical research at National Center for Agricultural Utilization Research, United States Department of Agriculture, Peoria, Illinois (2006). His research interests include high cell density culture, biodegradable polymers, polyhydroxyalkanoates, the biosynthesis and applications of nanomaterials, and biorefinery.

Arvind Gupta (Dr.) received his PhD in Chemical Engineering from the Indian Institute of Technology (IIT) Guwahati, India in 2017 where he studied stereocomplex poly(lactic acid) and its biocomposites. He then joined Chungbuk National University, Cheongju, Republic of Korea in July 2018 as postdoctoral fellow. Further, he joined University of Guelph in August 2019 as a postdoctoral fellow. His main research interests are polymer science, polymer characterization, biopolymers, nanocomposites, renewable resources, and eco-compatible materials.

nanomaterials

MDPI

Editorial

Soft, Biological and Composite Nanomaterials

Arvind Gupta and Beom Soo Kim *

Department of Chemical Engineering, Chungbuk National University, Cheongju, Chungbuk 28644, Korea; arvind@chungbuk.ac.kr

* Correspondence: bskim@chungbuk.ac.kr; Tel.: +82-43-261-2372

Received: 24 July 2020; Accepted: 28 July 2020; Published: 29 July 2020

The progress in the area of nanotechnology has opened the door for the fabrication of soft, biological and composite nanomaterials for targeted applications. Nanomaterials are known to enhance the properties and functionality of composite materials severalfold. The properties for the desired applications can often be achieved by the addition of small amounts of nanomaterials into soft materials such as polymers, gels, and biomaterials. Various techniques such as the functionalization of nanomaterials and the fabrication of composites in situ are ground-breaking methods that may lead to a significant improvement in the properties of these materials. Furthermore, there is a need for a focused characterization of the developed materials in order to use them for targeted applications, which will ultimately contribute to the future development of nanomaterials and their composites for various applications. Nanomaterials, such as nanoparticles and graphene, are also found to have a tremendous potential for a wide variety of biomedical applications such as antimicrobial and antitumor agents, drug delivery, tissue engineering, biosensors, bioimaging, and enzyme mimics. Therefore, there is a growing need to develop environmentally friendly processes for nanomaterial synthesis, such as biological methods using microorganisms, enzymes, and plants/plant extracts.

The current special issue of nanomaterials features an overview of several articles (eight research articles and one review article), wherein the use of various nanomaterials such as nanodiamonds, gold nanoparticles, nanochitosan, graphene oxide nanoparticles, and titanium nanofibers have been shown.

The use of nanodiamonds (NDs) for the fabrication of scaffolds may be effective in tissue regeneration. The inherent properties of NDs, such as a relatively low cytotoxicity, fluorescence, large specific surface area, and hardness, render them as potential nanomaterials for composite tissue engineering scaffolds. The incorporation of NDs (45 nm) as fillers in a well-known biocompatible polymer, i.e., poly(ε-caprolactone) (PCL) was investigated by Fox et al. [1] who observed that the composites possess biocompatibility and better degradation properties along with processability. The presence of NDs imparts roughness to the surface of PCL, thereby improving its hydrophilicity and promoting the degradation rate with a compromise on its tensile strength. An enhanced adhesion of osteoblast cells to the composite in comparison to the virgin PCL paves the way for the development of advanced tissue engineering scaffolds. Further, Nirwan et al. [2] fabricated a material using a neutralized chitosan/polyethylene oxide-based nanofiber decorated with gold nanoparticles for the development of tissue engineering scaffolds via an electrospinning technique. They proposed a simple neutralization process in order to conserve the structural integrity of the fabricated nanofibers over a period of six months. The neutralization of the reactive NH_3^+ group using potassium carbonate and transforming it into NH_2 makes this material insoluble in the biological system. This enhances the mechanical integrity of the product over six months in the biological solution and renders it suitable for tissue engineering applications.

Nanoceria as a nanomaterial can be used for the detection of galactose, which is an important marker for the diagnosis of galactosemia. Nguyen et al. [3] developed a composite of ceria oxide nanoparticles and galactose oxidase entrapped in an agarose gel, and this was found to be capable of detecting galactose. The enzymatic catalysis of galactose generates hydrogen peroxide (H_2O_2)

which induces a yellow color in the system without any chromogenic substrate. The calibration of the generated color using an electronic system can detect galactose with a sensitivity as low as 0.05 mM.

Further, graphene oxide (GO) which is known for its unique mechanical, thermal, optical, and electrical characteristics, is found to have a potential use in biomedical applications, apart from its widespread use in electronics and chemical applications. The different sizes of GO nanoparticles may have different effects on cell proliferation. Gurunathan et al. [4] conducted a study to understand the cyto- and geno-toxic effect of GO nanoparticles on germ (TM3 and TM4) cells. They concluded that both 20 nm and 100 nm sized GO nanoparticles exert a potent cytotoxic effect by reducing cell viability and proliferation. They found that the smaller sized (20 nm) GO has a more negative effect than 100 nm GO. Through this study, they revealed that GO is not 100% safe for biomedical applications and a thorough investigation is required, even though GO has the potential to be used for biomedical applications.

For other areas of biomedical applications where the shape memory of a polymer is predominantly required, Gupta et al. [5] developed polyurethane biocomposites. They found a way to utilize chitosan for the fabrication of biocomposites. Usually, chitosan is not compatible with hydrophobic polymers such as PCL. They transformed chitosan flakes to nanochitosan using a chemical treatment and incorporated them into the polyurethane matrix in situ in such a way that they can form a chemical linkage. The presence of nanochitosan in the polyurethane matrix acted as a chemical crosslinked node that governed the crystallinity of the biocomposite, thereby stimulating the shape memory with enhanced mechanical properties.

Another area of research in biomedical applications is drug delivery systems. The use of micro and nanostructures can be exploited for diagnostic and therapeutic applications. Nanobubbles, which can consist of a shell made of phospholipids, polymers, or proteins surrounding the core of a less soluble gas, are utilized for gas delivery applications. Khan et al. [6] aimed at identifying the role of the composition and use of polyethylene glycol (PEG) as a surfactant in the development of oxygen nanobubbles. They found that an increase in the content of PEG leads to a reduction in the nanobubble size and its distribution. The use of the generated oxygen nanobubbles was found to be non-toxic and did not cause hemolysis in sheep blood. Oxygen nanobubbles were also employed in an ultrasound imaging technique and were found to be traceable. Therefore, oxygen nanobubbles can be used for gas delivery applications and ultrasound imaging as well.

The controlled mechanism of the drug release rate is an important parameter in drug release or delivery systems. The use of magnetic nanoparticles can be effective in this application because it allows external stimulation. Mierzwa et al. [7] developed a biocompatible hybrid cubical phase nanomaterial called magnetocubosomes, containing hydrophobic magnetic nanoparticles which allow for the control of the release of the drug methotrexate at appropriate sites. It can be easily separated or relocated under the influence of magnetic fields.

Nanomaterials have become extensively important in photovoltaic applications as well. Titanium oxides (TiO_2) nanoparticles can be used in dye-sensitized solar cells (DSSCs) as a transparent conductive layer. Jo et al. [8] used an electrospinning technique to fabricate porous and dense TiO_2 nanofibers with 200 nm diameters as an additive in TiO_2 nanoparticles for DSSCs. They found that the incorporation of nanofibers enhances the performance of DSSCs by improving the charge transport and accessibility to electrolyte ions. It also enhanced the absorption of visible light in comparison to TiO_2 nanoparticles.

Furthermore, this special issue contains a review article by Perepelkin et al. [9] which focuses on different depth-sensing indentation (DSI) approaches and factors to characterize the mechanical properties of soft, biological, and biomimetic materials at the micro- and nanoscale.

Overall, this special issue highlights investigations in highly diverse research fields to encourage multifaceted research in the scientific community.

Author Contributions: A.G. wrote original manuscript. B.S.K. reviewed and revised manuscript. All authors have read and agreed to the published version of the manuscript.

Funding: This research was supported by the National Research Foundation of Korea (NRF-2019R 1I1A3A02058523).

Conflicts of Interest: The authors declare no conflict of interest.

References

1. Fox, K.; Ratwatte, R.; Booth, M.A.; Tran, H.M.; Tran, P.A. High Nanodiamond Content-PCL Composite for Tissue Engineering Scaffolds. *Nanomaterials* **2020**, *10*, 948. [CrossRef] [PubMed]
2. Nirwan, V.P.; Al-Kattan, A.; Fahmi, A.; Kabashin, A.V. Fabrication of Stable Nanofiber Matrices for Tissue Engineering via Electrospinning of Bare Laser-Synthesized Au Nanoparticles in Solutions of High Molecular Weight Chitosan. *Nanomaterials* **2019**, *9*, 1058. [CrossRef] [PubMed]
3. Nguyen, P.T.; Ahn, H.T.; Kim, M.I. Reagent-Free Colorimetric Assay for Galactose Using Agarose Gel Entrapping Nanoceria and Galactose Oxidase. *Nanomaterials* **2020**, *10*, 895. [CrossRef] [PubMed]
4. Gurunathan, S.; Kang, M.H.; Jeyaraj, M.; Kim, J.H. Differential Cytotoxicity of Different Sizes of Graphene Oxide Nanoparticles in Leydig (TM3) and Sertoli (TM4) Cells. *Nanomaterials* **2019**, *9*, 139. [CrossRef] [PubMed]
5. Gupta, A.; Kim, B.S. Shape Memory Polyurethane Biocomposites Based on Toughened Polycaprolactone Promoted by Nano-Chitosan. *Nanomaterials* **2019**, *9*, 225. [CrossRef] [PubMed]
6. Khan, M.S.; Hwang, J.; Lee, K.; Choi, Y.; Jang, J.; Kwon, Y.; Hong, J.W.; Choi, J. Surface Composition and Preparation Method for Oxygen Nanobubbles for Drug Delivery and Ultrasound Imaging Applications. *Nanomaterials* **2019**, *9*, 48. [CrossRef] [PubMed]
7. Mierzwa, M.; Cytryniak, A.; Krysiński, P.; Bilewicz, R. Lipidic Liquid Crystalline Cubic Phases and Magnetocubosomes as Methotrexate Carriers. *Nanomaterials* **2019**, *9*, 636. [CrossRef] [PubMed]
8. Jo, M.S.; Cho, J.S.; Wang, X.L.; Jin, E.M.; Jeong, S.M.; Kang, D.W. Improving of the Photovoltaic Characteristics of Dye-Sensitized Solar Cells Using a Photoelectrode with Electrospun Porous TiO_2 Nanofibers. *Nanomaterials* **2019**, *9*, 95. [CrossRef] [PubMed]
9. Perepelkin, N.V.; Borodich, F.M.; Kovalev, A.E.; Gorb, S.N. Depth-Sensing Indentation as a Micro- and Nanomechanical Approach to Characterisation of Mechanical Properties of Soft, Biological, and Biomimetic Materials. *Nanomaterials* **2020**, *10*, 15. [CrossRef]

nanomaterials

MDPI

Article

High Nanodiamond Content-PCL Composite for Tissue Engineering Scaffolds

Kate Fox [1,*], Rahul Ratwatte [1], Marsilea A. Booth [1], Hoai My Tran [2,3] and Phong A. Tran [2,3,*]

1 Center for Additive Manufacturing, School of Engineering, RMIT University, Melbourne, VIC 3000, Australia; rahul.ratwatte@unimelb.edu.au (R.R.); marsilea.harrison@rmit.edu.au (M.A.B.)

2 Interface science and materials engineering group, School of Mechanical, Medical and Process Engineering, Queensland University of Technology (QUT), 2 George Street, Brisbane QLD 4000, Australia; hoaimy.tran@qut.edu.au

3 Institute of Health and Biomedical Innovation, Queensland University of Technology, Kelvin Grove, QLD 4059, Australia

* Correspondence: kate.fox@rmit.edu.au (K.F.); phong.tran@qut.edu.au (P.A.T.)

Received: 26 April 2020; Accepted: 13 May 2020; Published: 15 May 2020

Abstract: Multifunctional scaffolds are becoming increasingly important in the field of tissue engineering. In this research, a composite material is developed using polycaprolactone (PCL) and detonation nanodiamond (ND) to take advantage of the unique properties of ND and the biodegradability of PCL polymer. Different ND loading concentrations are investigated, and the physicochemical properties of the composites are characterized. ND-PCL composite films show a higher surface roughness and hydrophilicity than PCL alone, with a slight decrease in tensile strength and a significant increase in degradation. Higher loading of ND also shows a higher osteoblast adhesion than the PCL alone sample. Finally, we show that the ND-PCL composites are successfully extruded to create a 3D scaffold demonstrating their potential as a composite material for tissue regeneration.

Keywords: nanodiamond; polycaprolactone; composite; 3D-printed scaffold

1. Introduction

There is a growing need for effective scaffolds for tissue regeneration in biomaterials research, with requirements including biocompatibility, strength, and structure. Detonation nanodiamonds (NDs) are gaining significant interest for their mechanical strength, optical properties, and biocompatibility [1–4]. NDs have been assessed for biocompatibility both in vitro and in vivo with positive outcomes such as low cytotoxicity [5–7], improved cellular adhesion [8], and improved cell proliferation [3]. They are significantly less cytotoxic than other carbon-based nanoparticles such as CNTs [7], and endocytic NDs are non-cytotoxic during cell division and differentiation [9]. Further to their biocompatibility, NDs exhibit other beneficial properties. Fluorescence in ND particles allows for non-invasive fluorescence tracking in both cells [10] and tissue [4]. The large specific surface area and unique surface structure of NDs are used to make drug delivery systems [11–13]. Antibacterial properties are also displayed by NDs, mainly due to surface chemical terminations [14–17]. Furthermore, NDs have been shown to prevent biofilm formation, particularly when combined with carbohydrates [18,19]. The low toxicity, improved cellular interactions, and biocompatibility make them an attractive material for composite tissue engineering scaffolds.

The structure of NDs consists of a diamond-like core with a graphitic outer shell that contains many oxygen-rich functional groups, attractive for developing polymer composites [20]. The addition of NDs into a polymeric matrix has been shown to enhance mechanical properties [21], including hardness and elastic modulus [22], abrasion and scratch resistance in an epoxy polymer matrix [23], and compressive strength in a poly(vinyl alcohol) matrix [22]. Polycaprolactone (PCL) is an FDA-approved polyester widely studied for soft and hard tissue engineering scaffolds because of its non-toxic nature, degradability, and low melting temperature [24]. Despite those excellent properties, this polymer still possesses limitations, including a slow degradation time (2–3 years in interstitial fluid [25]), low mechanical strength, and high hydrophobicity [26,27], which hinder cell interactions [3,4,25]. Co-polymers and blended polymer composites are shown to improve PCL properties, particularly to tailor degradation properties [28]. A composite of PCL, poly(lactide-*co*-glycolide) (PLGA) and tricalcium phosphate showed a faster degradation speed [29], while a co-polymer of PCL and δ-valerolacton also exhibited faster degradation rates than PCL alone [30]. A review by Bartnikowski et al. covers different PCL degradation mechanisms within physiological contexts in detail [28]. Modification of PCL, as an ideal biomaterial for scaffold design, thus involves reduction of its hydrophobicity and modification of degradation rate to be in harmony with tissue regeneration rate [25].

In this study, we use NDs in a PCL matrix in order to make a biodegradable composite (ND-PCL) with improved cellular interactions. This builds upon the previous work within our group in which we formed a 0.1% wt ND-PCL composite [4]. Here we extend the ND loading to 10% wt and 20% wt and determine that ND incorporation affects the physicochemical properties of PCL, namely the mechanical and biodegradation properties of the ND-PCL composite. Lastly, we use osteoblasts to investigate the cellular response to the composite. We chose osteoblasts as a relevant cell type since the composite scaffolds we form herein are aimed for bone regeneration e.g., maxillofacial surgical implants. This builds upon more traditional ND-PCL film fabrication methods such as evaporation giving free-standing thin films [4] and electrospinning [3,21,31]. Herein we instead use additive manufacturing thermal extrusion-based 3D printing to show spatial control of the ND-PCL composites. Additive manufacturing techniques allow for design tailoring [25,32], scaffold porosity, and personalized patient care [33]. We show that ND-PCL composites have great potential as scaffolds for tissue engineering, displaying good biocompatibility, degradation properties and processability.

2. Materials and Methods

2.1. Fabrication of ND-PCL Composites

Fabrication of ND-PCL composites. The PCL (Mn = 80,000) was from Sigma Aldrich (Castle Hill, NSW, Australia); NDs of 45 nm were obtained from NaBond (Nabond Technologies, China) and irradiated as detailed in [34]. In brief, as-received ND were dispersed in deionized water at a concentration of 1 mg mL^{-1}. Centrifugation was used to remove large aggregates prior to irradiation with high energy electrons (2 MeV to a total fluence of 1×10^{18} cm^{-2}). The ND material was then annealed in a vacuum at 800 °C for 2 h in order to induce vacancy diffusion and ND formation. ND-PCL composites were produced as films for characterization, biological compatibility testing and additive manufacturing. The fabrication process involved physical blending of ND and PCL at 5%, 10% and 20% ND wt% in trichloromethane. The mixture suspension was cast onto glass dishes and free-standing composite films (~0.1 mm) were removed after complete solvent evaporation.

2.2. Characterization of ND-PCL Composites

Physicochemical characterization. Scanning electron microscope images were used to determine sample morphology. Samples were studied under the scanning electron microscope (FET Quanta ESEM (Thermo Fisher, OR, USA) 30 kV accelerating voltage, with a working distance of 10.6 mm, spot size 5 in variable (VP) pressure mode). Static water contact angle measurements were used [35] for 3 samples per condition and 3 repeats per sample, with the results then averaged. Samples (5 mm Ø) were pre-treated via immersion in ethanol 80% for 2 h and left to evaporate until completely dry, and mounted onto glass slides for examination [36]. Fourier-transformed infrared spectroscopy analysis was performed on a Nicolet FTIR spectrophotometer (Nicolet Analytical Instruments (Thermo Fisher, OR, USA)) using a setting of 64 scan-average and a resolution of 1 cm^{-1}. Thermal analysis differential scanning calorimetry (TA Instrument, DSC Q100 (Rydalmere NSW, Australia)) was used to measure the melting temperature (Tm) in order to evaluate the crystallinity of the polymer. Pure PCL and treated PCL samples (4–5 mg) were heated at rate of 10 °C/min from −60 °C to 100 °C, and the crystallinity calculated based on the equation [37,38]:

$$\chi(\%) = \left(\frac{H_{sample}}{H_0}\right) \times 100 \tag{1}$$

where: $\chi(\%)$ is the crystallinity percentage, H_{sample} is the sample's enthalpy of fusion and H_0 is the enthalpy of fusion for a 100% crystalline PLC, with a value of 136.1 J/g.

Tensile strength measurement. The mechanical properties of the films were examined using an Instron 4302 Material Testing System operated by Series IX Automated Materials Tester version 7.43 system software with a 1 kN load-cell. Samples were cut into dog bone shapes (12.7 mm width, 38 mm length in the middle section, 79 mm total length) in accordance with the ASTM D695-96 guidelines and subject to elongation at a rate of 1 mm/min until failure.

Accelerated in vitro degradation. Sodium hydroxide (2 M) was used to accelerate the hydrolysis reaction [39]. Replicates of 14.5 to 16.0 mg pieces of PCL and ND-PCL films (24) were submerged in 2.0 mL NaOH in closed Eppendorf tubes, and maintained at 37 °C. At time points (12, 24, 48 and 72 h) the films were removed and rinsed thoroughly with de-ionized water. The samples were then dried, placed in an oven at 35 °C for 48 h and weighed to calculate the percentage mass loss.

2.3. *Investigating Cellular Interactions with ND-PCL Composites and Performing Additive Manufacturing*

Cell culture. Cell culture was performed using primary human osteoblasts (between 2–6 passages) cultured in DMEM supplemented with 10% fetal bovine serum and 1% penicillin/streptomycin. For the adhesion assay, cells were seeded onto 5 mm Ø film samples at a density of 800 cells/sample and incubated for 4 h. Cells were fixed in a 4% formaldehyde, permeabilised with 0.2% Triton X-100/PBS, stained with 0.5% BSA/PBS containing 0.8 ug/mL TRITC-conjugated phalloidin and 5 µg/mL DAPI, and imaged.

Additive Manufacturing. The 20 wt% ND-PCL composite was used to make 3D scaffolds using a layer-by-layer melt-screw extrusion through a 20-gauge needle at an extrusion temperature of 175 °C. The composites are printed using a custom-made bioextruder.

2.4. *Statistical Analysis*

The results are reported as mean and standard deviations and a student *t*-test is used to analyze statistical significance between means.

3. Results

3.1. *Physicochemical Properties and Characterization of ND-PCL Composites*

Figure 1A–C shows the surface morphology at different ND loading conditions as investigated by scanning electron microscopy. PCL samples showed a smooth surface with large and uneven domains, whilst the addition of NDs resulted in a change in surface topology. The addition of 10% ND to PCL resulted in surface particle formation, most likely ND clusters. Meanwhile, at 20% ND loading a rougher surface with several granular clusters could be seen. The addition of 20% ND to the PCL showed a significant ($p \leq 0.05$) increase in the wettability of the films, while the 10% ND-PCL composite film did not show a large difference from PCL film wettability (Figure 1D). The material tensile properties decreased as the ND *w/w* fraction increased (Figure 1E,F). The elastic modulus of both composite conditions produced lower stiffness compared to PCL, with the 20% *w/w* ND-PCL composite closest to that of PCL.

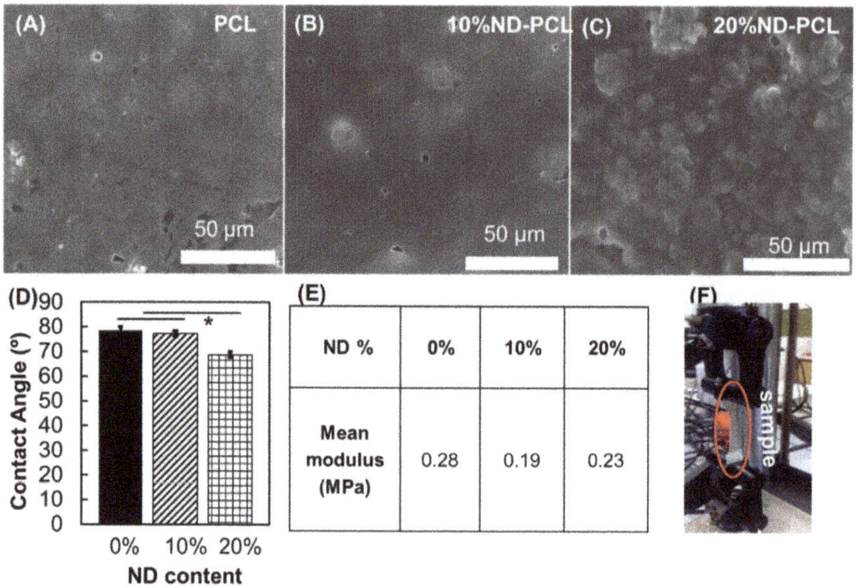

Figure 1. (**A–C**) SEM surface imaging of (**A**) polycaprolactone (PCL) (**B**) nanodiamond-polycaprolactone (ND-PCL) films with 10% ND loading and (**C**) 20% ND loading. (**D**) Contact angle measurement (n = 3, student *t*-test ($p \leq 0.05$)). (**E**) Mean tensile modulus for films with different ND loadings ($n \geq 6$) and (**F**) setup of the tensile test.

Differential scanning calorimetry showed that the crystallinity of PCL decreased as the ND content increased. PCL had a percentage crystallinity of 64% which reduced to 62% for 10% ND-PCL and to 57% for 20% ND-PCL (Figure 2A,B). However, both ND-PCL composite samples showed significant degradation over the observed time period (Figure 2C). The 20% ND-PCL composite showed an initial loss of ~40% mass, followed by a slowed loss before a final mass loss of 74% after 70 h. FTIR spectra showed no significant difference in chemical bonding, indicating that the interaction between ND and PCL uses physical bonds (Figure 2D).

3.2. *Cellular Interactions with ND-PCL Composites and Additive Manufacturing of Composites*

A cell adhesion assay was used to determine the surface interactions between human osteoblasts and the composite ND-PCL samples. Figure 3A shows a marked improvement in the attachment of human osteoblast cells after the 10% and 20 wt% ND composites are added to the PCL films (Figure 3A). The cells also appeared to remain viable on the composite materials after 14 days in culture (Figure S1). A proof of concept printing trial was performed to investigate whether including ND into PCL enables extrusion and additive manufacturing of ND-PCL composites. Figure 3B shows the results of an extruded 20% ND loaded composite film. The extruded ND-PCL composite appeared to hold its shape with no limitation in flow of the composite through the extrusion nozzle (Figure 3B).

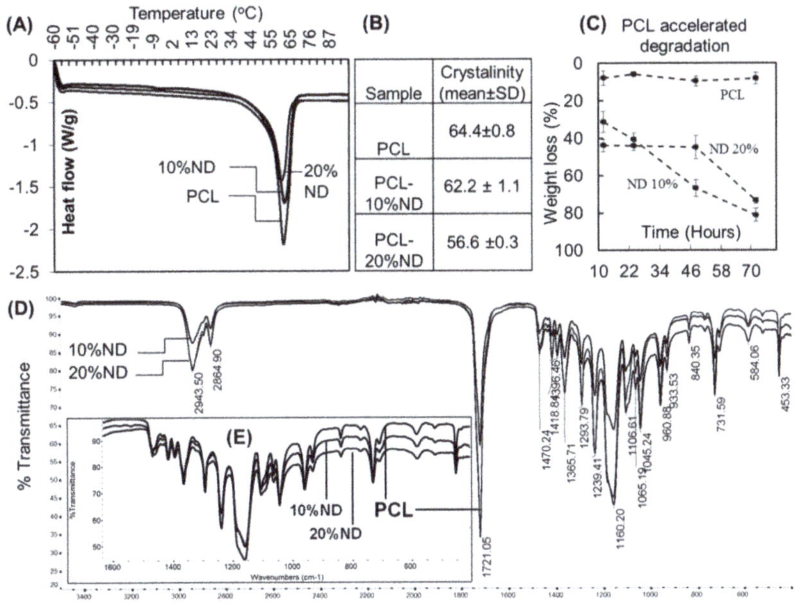

Figure 2. (**A**,**B**) Crystallinity of PCL and ND-PCL composites and their representative differential scanning calorimetry (DSC) traces (n = 5). (**C**) Weight loss of samples in accelerated degradation experiments (n = 6). (**D**) Representative FTIR spectra of PCL and ND-PCL composites. (**E**) Inset showing FTIR spectra in the 400–1600 cm^{-1}.

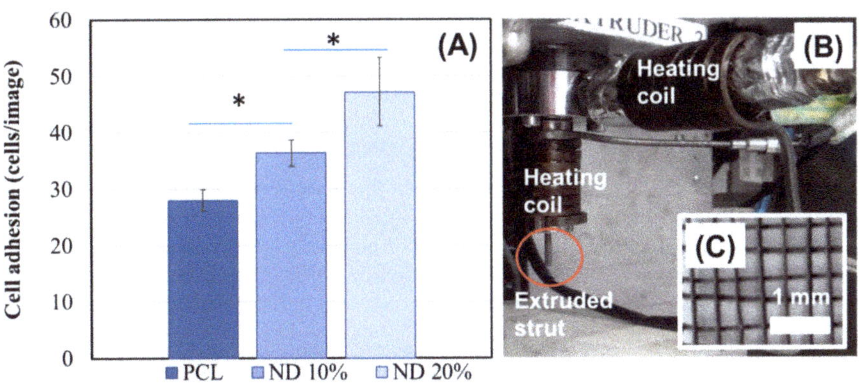

Figure 3. (**A**) Osteoblast adhesion on composite materials showing higher adhesion compared to PCL alone (mean ± S.E.M, n = 6, student *t*-test ($p \leq 0.05$)). (**B**) Proof of concept of the 3D printing compatibility of PCL-ND 20% composite showing the melt-extrusion setup and (**C**) a scaffold printed using layer-by-layer deposition of extruded struts (nozzle size: 20-gauge needle).

4. Discussion

4.1. Inclusion of ND into a PCL Scaffold Modifies the Material Surface

As the interest in tissue engineering increases new materials are required to improve the interface. Here we show a scaffold composed of polycaprolactone and detonation nanodiamonds, ND-PCL, with ND inclusion of both 10% ND-PCL composite and 20% ND-PCL. By adding ND into the PCL, the newly formed composite material reports a lower tensile strength and decrease in crystallinity coupled with faster degradation when we compare the two composites to the PCL alone. Further, the degradation profiles of the 10% and 20% ND-PCL composites are different with the 20% ND composite able to resist dissolution for a longer time period than the 10% ND composite. This is because NDs are likely to supply additional surface nucleation sites during film drying, increasing the surface roughness and hence the hydrophilicity of the material. As expected there is a limited mass decrease that occurs for PCL during the 70 h exposure to NaOH [40]. Increased and tunable degradation profiles are useful for implantable biomaterials, where material degradation occurs ideally in a controlled manner. The chosen biomaterial should degrade at a rate aligned with the rate of tissue regeneration [25]. Our results show that a degradation rate can be tailored by using composite ND-loading to meet requirements. Our previous results show ND-PCL composites exhibit the potential for tracking degradation in situ via sub-dermal fluorescent imaging [4]. The ability to track and even tune a timely degradation of material is highly coveted for biomedical implant scaffolds.

4.2. Inclusion of ND into a PCL Scaffold Modifies Biointerface

As ND is incorporated into the PCL material, the biointerface appears to be more supportive of osteoblast adhesion, with a clear improvement in cell attachment with increased ND loads. Osteoblasts are an established key in the bone regeneration cycle and as such it is important that the new composite material can support their attachment and proliferation. The improvement in adhesion with 20% ND compared to the PCL is likely linked to the improved hydrophilicity of the ND-PCL scaffold. Hydrophilicity is an important property for scaffold biomaterials. This result supports our previous work [4] where ND-PCL was found to be a superior scaffold for fibroblast attachment, and that of others [3,25] who observe that improvements in PCL-composite hydrophilicity by adding ND can equate to better cellular adhesion. This is also the case with osteoblasts as well as fibroblasts [4,25] and Chinese hamster ovarian (CHO) cells [3]. The biointerface shows an increased surface roughness as ND content increases, contributing to the increased hydrophilicity. This supports the finding of Jeon et al. [41], who used oxygen plasma treatment on PCL scaffolds to tailor a range of surface roughness topologies, that the surface roughness enhanced initial cell adhesion [41]. The increase in roughness of our ND-PCL composites as compared to PCL is a likely contributor to the improved osteoblast cell adhesion observed and a vital first step towards promoting osseointegration.

4.3. Fabrication and Additive Manufacturing of ND-PCL Scaffolds

The fabrication of ND-PCL composites is limited by the fabrication techniques used to manufacture the composites. Using casting and electrospinning, significantly lower ND concentrations of 0.1–6% ND *w/v* [3,4,21,42] are used to make composites, while herein we are increasing this to 20% ND *w/w* loading. Here, we use the ND-PCL composite as a base material in a custom-made bioextruder for additive manufacturing. Following optimization, it was found that the 20% ND-PCL blend can be effectively extruded through a 20-gauge needle (internal diameter of 603 μm [43]), providing the ability to fabricate customized scaffold designs. A simple design was used as a prototype scaffold (Figure 3C); however, the 3D printing of scaffold design has powerful capabilities. This increase in ND wt% produces beneficial changes in the physical properties of the scaffold; its degradation profile and its hydrophilicity, thereby improving the biointerfacial cellular interactions. Since the aim is to use these scaffolds for tissue engineering, the higher ND content should also be considered in terms of toxicity and clearance. Although a higher concentration of ND is used, the size of a scaffold is likely to be small. However it is important to note that NDs are shown to have high biocompatibility and low toxicity [5–7,9]. However, clearance is an important issue to consider. Fortunately, the inherent fluorescence in NDs can be used to investigate in situ degradation of the scaffold as highlighted by our previous work [4], and in vivo imaging [10]. Composites loaded with other diamond structures [44], or other base polymers [31] have incorporated higher material loadings, with other applications in mind and without the beneficial properties of NDs or PCL.

When designing biomedical tissue engineering scaffolds, it is best to control morphology at multiple structural levels to meet clinical requirements [25]. This can be thought of in two scales, macroscale and microscale. Macroscale can include the external architecture of the implant, mechanical properties, and scaffold density. Microscale, on the other hand, can refer to the material porosity, surface topology, and the degradation capabilities of the material. Changes to processing parameters can be used in additive manufacturing to tailor PCL scaffolds [25,32,39]. Additive manufacturing can contribute to control of the microscale via porosity of the printed scaffold. Large-scale porosity can be adjusted depending on printing conditions and design. A balance needs to be achieved to maintain advantageous mechanical properties while promoting osteoblast migration, integration, nutrient transfer, and vascularization. Regeneration occurs as cells grow either within the scaffold itself or shifting from neighboring tissue, highlighting the importance of the scaffold microstructure. On the macroscale, additive manufacturing can greatly improve the ease of design, precision, resolution, and individualization of biomedical implants. 3D printing coupled to 3D scanning can potentially offer solutions for patient-specific care [33], with implants able to match shape requirements and, therefore, improve implant success.

Our future work based on this current proof-of-principle study will focus on investigating more complex designs for 3D printing, and subsequently test these complex designs for their physicochemical properties and cellular interactions. Characterization of the 3D printed structures must consider printed structure, surface roughness, contact angle measurement, mechanical characterization, and degradation studies. Mechanical properties and mass loss are strongly influenced by construct geometry; therefore the degradation rate of 3D structures may show interesting behavior [28]. Coatings and additions to improve scaffold use for bone regeneration may also be considered [45,46]. Subsequent steps include cellular interactions of ND-PCL 3D printed composites with osteoblasts and finally, in vivo testing of promising composites. The results from this study highlight the exciting potential of ND-PCL composites. By combining fluorescent degradation tracking, tunable degradation profile, improved surface wettability, and additive manufacturing capacity, ND-PCL composites have high potential for tissue regeneration scaffolds.

5. Conclusions

Here, we investigated the potential of ND-PCL composites as a biomaterial for tissue engineering scaffolds. Our findings showed that high loading of ND in PCL was possible (up to 20% *w*/*w*), and this changed the physicochemical properties of the composite. Tensile properties decreased slightly, while a marked increase in degradation was observed after ND incorporation. The hydrophilicity of the composite was greatly increased after ND addition, likely a feature of surface roughness. This, in turn, contributes to the increased adhesion of osteoblast cells observed on ND-PCL composites as compared to PCL. Lastly, ND-PCL composite could be additively manufactured into 3D scaffolds via melt-extrusion, paving the way for preparation of advanced tissue engineering scaffolds.

Supplementary Materials: The following are available online at http://www.mdpi.com/2079-4991/10/5/948/s1, Figure S1: Representative images of cells cultured on PCL, ND-PCL 10% and ND-PCL 20% for 2 weeks.

Author Contributions: Conceptualization, P.A.T. and K.F.; methodology, R.R. and P.A.T.; formal analysis, H.M.T., R.R. and P.A.T.; writing—original draft preparation, H.M.T., R.R. and M.A.B.; writing—review and editing, M.A.B., K.F. and P.A.T.; supervision, K.F. and P.A.T.; funding acquisition, K.F. All authors have read and agreed to the published version of the manuscript

Funding: K.F. is supported by the Clive and Vera Ramaciotti Foundation and the CASS Foundation (Medicine and Science). P.A.T. is supported by Advance Queensland Research Fellowship (AQRF04816-17RD2) and QUT's Vice Chancellor Research Fellowship.

Acknowledgments: The authors acknowledge the facilities and technical assistance of the RMIT Advanced Manufacturing Precinct. The authors acknowledge the facilities, and the scientific and technical assistance, of the Australian Microscopy & Microanalysis Research Facility at RMIT University.

Conflicts of Interest: The authors declare no conflict of interest.

References

1. Whitlow, J.; Pacelli, S.; Paul, A. Multifunctional nanodiamonds in regenerative medicine: Recent advances and future directions. *J. Control Release* **2017**, *261*, 62–86. [CrossRef] [PubMed]
2. Schrand, A.M.; Dai, L.; Schlager, J.J.; Hussain, S.M.; Osawa, E. Differential biocompatibility of carbon nanotubes and nanodiamonds. *Diam. Relat. Mater.* **2007**. [CrossRef]
3. Houshyar, S.; Kumar, G.S.; Rifai, A.; Tran, N.; Nayak, R.; Shanks, R.A.; Padhye, R.; Fox, K.; Bhattacharyya, A. Nanodiamond/poly-ε-caprolactone nanofibrous scaffold for wound management. *Mater. Sci. Eng. C* **2019**, *100*, 378–387. [CrossRef] [PubMed]
4. Fox, K.; Tran, P.A.; Lau, D.W.M.; Ohshima, T.; Greentree, A.D.; Gibson, B.C. Nanodiamond-polycaprolactone composite: A new material for tissue engineering with sub-dermal imaging capabilities. *Mater. Lett.* **2016**, *185*, 185–188. [CrossRef]
5. Yuan, Y.; Wang, X.; Jia, G.; Liu, J.H.; Wang, T.; Gu, Y.; Yang, S.T.; Zhen, S.; Wang, H.; Liu, Y. Pulmonary toxicity and translocation of nanodiamonds in mice. *Diam. Relat. Mater.* **2010**, *19*, 291–299. [CrossRef]
6. Schrand, A.M.; Huang, H.; Carlson, C.; Schlager, J.J.; Osawa, E.; Hussain, S.M.; Dai, L. Are diamond nanoparticles cytotoxic? *J. Phys. Chem. B* **2007**, *111*, 2–7. [CrossRef]
7. Rifai, A.; Pirogova, E.; Fox, K. Diamond, Carbon Nanotubes and Graphene for Biomedical Applications. *Encycl. Biomed. Eng.* **2019**, 97–107. [CrossRef]
8. Houshyar, S.; Sarker, A.; Jadhav, A.; Kumar, G.S.; Bhattacharyya, A.; Nayak, R.; Shanks, R.A.; Saha, T.; Rifai, A.; Padhye, R.; et al. Polypropylene-nanodiamond composite for hernia mesh. *Mater. Sci. Eng. C* **2020**, *111*, 110–780. [CrossRef]
9. Liu, K.K.; Wang, C.C.; Cheng, C.L.; Chao, J.I. Endocytic carboxylated nanodiamond for the labeling and tracking of cell division and differentiation in cancer and stem cells. *Biomaterials* **2009**, *30*, 4249–4259. [CrossRef]
10. Chang, Y.R.; Lee, H.Y.; Chen, K.; Chang, C.C.; Tsai, D.S.; Fu, C.C.; Lim, T.S.; Tzeng, Y.K.; Fang, C.Y.; Han, C.C.; et al. Mass production and dynamic imaging of fluorescent nanodiamonds. *Nat. Nanotechnol.* **2008**, *3*, 284–288. [CrossRef]
11. Li, J.; Zhu, Y.; Li, W.; Zhang, X.; Peng, Y.; Huang, Q. Nanodiamonds as intracellular transporters of chemotherapeutic drug. *Biomaterials* **2010**, *31*, 8410–8418. [CrossRef] [PubMed]
12. Zhu, Y.; Li, J.; Li, W.; Zhang, Y.; Yang, X.; Chen, N.; Sun, Y.; Zhao, Y.; Fan, C.; Huang, Q. The biocompatibility of nanodiamonds and their application in drug delivery systems. *Theranostics* **2012**, *2*, 302–312. [CrossRef] [PubMed]
13. Shimkunas, R.A.; Robinson, E.; Lam, R.; Lu, S.; Xu, X.; Zhang, X.Q.; Huang, H.; Osawa, E.; Ho, D. Nanodiamond-insulin complexes as pH-dependent protein delivery vehicles. *Biomaterials* **2009**, *30*, 5720–5728. [CrossRef] [PubMed]
14. Wehling, J.; Dringen, R.; Zare, R.N.; Maas, M.; Rezwan, K. Bactericidal activity of partially oxidized nanodiamonds. *ACS Nano* **2014**, *8*, 6475–6483. [CrossRef] [PubMed]
15. Rifai, A.; Tran, N.; Reineck, P.; Elbourne, A.; Mayes, E.; Sarker, A.; Dekiwadia, C.; Ivanova, E.P.; Crawford, R.J.; Ohshima, T.; et al. Engineering the Interface: Nanodiamond Coating on 3D-Printed Titanium Promotes Mammalian Cell Growth and Inhibits Staphylococcus aureus Colonization. *ACS Appl. Mater. Interfaces* **2019**, *11*, 24588–24597. [CrossRef] [PubMed]
16. Beranová, J.; Seydlová, G.; Kozak, H.; Potocký, Š.; Konopásek, I.; Kromka, A. Antibacterial behavior of diamond nanoparticles against *Escherichia coli*. *Phys. Status Solidi* **2012**, *249*, 2581–2584. [CrossRef]
17. Jira, J.; Rezek, B.; Kriha, V.; Artemenko, A.; Matolínová, I.; Skakalova, V.; Stenclova, P.; Kromka, A. Inhibition of e. Coli growth by nanodiamond and graphene oxide enhanced by luria-bertani medium. *Nanomaterials* **2018**, *8*, 140. [CrossRef]
18. Barras, A.; Martin, F.A.; Bande, O.; Baumann, J.S.; Ghigo, J.M.; Boukherroub, R.; Beloin, C.; Siriwardena, A.; Szunerits, S. Glycan-functionalized diamond nanoparticles as potent E. coli anti-adhesives. *Nanoscale* **2013**, *5*, 2307–2316. [CrossRef]
19. Fessele, C.; Wachtler, S.; Chandrasekaran, V.; Stiller, C.; Lindhorst, T.K.; Krueger, A. Thiourea-Bridged Nanodiamond Glycoconjugates as Inhibitors of Bacterial Adhesion. *Eur. J. Org. Chem.* **2015**, *2015*, 5519–5525. [CrossRef]

20. Karami, P.; Salkhi Khasraghi, S.; Hashemi, M.; Rabiei, S.; Shojaei, A. Polymer/nanodiamond composites—A comprehensive review from synthesis and fabrication to properties and applications. *Adv. Colloid Interface Sci.* **2019**, *269*, 122–151. [CrossRef]
21. Ahn, G.Y.; Ryu, T.K.; Choi, Y.R.; Park, J.R.; Lee, M.J.; Choi, S.W. Fabrication and optimization of Nanodiamonds-composited poly(ε-caprolactone) fibrous matrices for potential regeneration of hard tissues. *Biomater. Res.* **2018**, *22*, 1–8. [CrossRef]
22. Maitra, U.; Prasad, K.E.; Ramamurty, U.; Rao, C.N.R. Mechanical properties of nanodiamond-reinforced polymer-matrix composites. *Solid State Commun.* **2009**, *149*, 1693–1697. [CrossRef]
23. Neitzel, I.; Mochalin, V.; Knoke, I.; Palmese, G.R.; Gogotsi, Y. Mechanical properties of epoxy composites with high contents of nanodiamond. *Compos. Sci. Technol.* **2011**, *71*, 710–716. [CrossRef]
24. Dorozhkin, S. Calcium Orthophosphate-Containing Biocomposites and Hybrid Biomaterials for Biomedical Applications. *J. Funct. Biomater.* **2015**, *6*, 708–832. [CrossRef] [PubMed]
25. Malikmammadov, E.; Tanir, T.E.; Kiziltay, A.; Hasirci, V.; Hasirci, N. PCL and PCL-based materials in biomedical applications. *J. Biomater. Sci. Polym. Ed.* **2018**, *29*, 863–893. [CrossRef] [PubMed]
26. Yildirim, E.D.; Pappas, D.; Güçeri, S.; Sun, W. Enhanced Cellular Functions on Polycaprolactone Tissue Scaffolds by O_2 Plasma Surface Modification. *Plasma Process. Polym.* **2011**, *8*, 256–267. [CrossRef]
27. Mattioli-Belmonte, M.; Vozzi, G.; Whulanza, Y.; Seggiani, M.; Fantauzzi, V.; Orsini, G.; Ahluwalia, A. Tuning polycaprolactone-carbon nanotube composites for bone tissue engineering scaffolds. *Mater. Sci. Eng. C* **2012**, *32*, 152–159. [CrossRef]
28. Bartnikowski, M.; Dargaville, T.R.; Ivanovski, S.; Hutmacher, D.W. Degradation mechanisms of polycaprolactone in the context of chemistry, geometry and environment. *Prog. Polym. Sci.* **2019**, *96*, 1–20. [CrossRef]
29. Idaszek, J.; Brynk, T.; Jaroszewicz, J.; Vanmeert, F.; Bruinink, A.; Święszkowski, W. Investigation of mechanical properties of porous composite scaffolds with tailorable degradation kinetics after in vitro degradation using digital image correlation. *Polym. Compos.* **2017**, *38*, 2402–2410. [CrossRef]
30. Fernández, J.; Etxeberria, A.; Sarasua, J.R. In vitro degradation studies and mechanical behavior of poly(ε-caprolactone-co-δ-valerolactone) and poly(ε-caprolactone-co-L-lactide) with random and semi-alternating chain microstructures. *Eur. Polym. J.* **2015**, *71*, 585–595. [CrossRef]
31. Behler, K.D.; Stravato, A.; Mochalin, V.; Korneva, G.; Yushin, G.; Gogotsi, Y. Nanodiamond-polymer composite fibers and coatings. *ACS Nano* **2009**, *3*, 363–369. [CrossRef] [PubMed]
32. Mota, C.; Puppi, D.; Dinucci, D.; Gazzarri, M.; Chiellini, F. Additive manufacturing of star poly(ε-caprolactone) wet-spun scaffolds for bone tissue engineering applications. *J. Bioact. Compat. Polym.* **2013**, *28*, 320–340. [CrossRef]
33. Muwaffak, Z.; Goyanes, A.; Clark, V.; Basit, A.W.; Hilton, S.T.; Gaisford, S. Patient-specific 3D scanned and 3D printed antimicrobial polycaprolactone wound dressings. *Int. J. Pharm.* **2017**, *527*, 161–170. [CrossRef] [PubMed]
34. Reineck, P.; Capelli, M.; Lau, D.W.M.; Jeske, J.; Field, M.R.; Ohshima, T.; Greentree, A.D.; Gibson, B.C. Bright and photostable nitrogen-vacancy fluorescence from unprocessed detonation nanodiamond. *Nanoscale* **2017**, *9*, 497–502. [CrossRef]
35. Bracco, G.; Holst, B. Surface science techniques. *Springer Ser. Surf. Sci.* **2013**, *51*. [CrossRef]
36. Yoganarasimha, S.; Trahan, W.R.; Best, A.M.; Bowlin, G.L.; Kitten, T.O.; Moon, P.C.; Madurantakam, P.A. Peracetic acid: A practical agent for sterilizing heat-labile polymeric tissue-engineering scaffolds. *Tissue Eng. Part. C Methods* **2014**, *20*, 714–723. [CrossRef]
37. Urbanczyk, L.; Calberg, C.; Stassin, F.; Alexandre, M.; Jérôme, R.; Jérôme, C.; Detrembleur, C. Synthesis of PCL/clay masterbatches in supercritical carbon dioxide. *Polymer (Guildf)* **2008**, *49*, 3979–3986. [CrossRef]
38. Benedict, C.V.; Cook, W.J.; Jarrett, P.; Cameron, J.A.; Huang, S.J.; Bell, J.P. Fungal degradation of polycaprolactones. *J. Appl. Polym. Sci.* **1983**, *28*, 327–334. [CrossRef]
39. Lee, S.H.; Lee, J.H.; Cho, Y.S. Analysis of degradation rate for dimensionless surface area of well-interconnected PCL scaffold via in-vitro accelerated degradation experiment. *Tissue Eng. Regen. Med.* **2014**, *11*, 446–452. [CrossRef]
40. Lam, C.X.F.; Savalani, M.M.; Teoh, S.H.; Hutmacher, D.W. Dynamics of in vitro polymer degradation of polycaprolactone-based scaffolds: Accelerated versus simulated physiological conditions. *Biomed. Mater.* **2008**, *3*. [CrossRef]

41. Jeon, H.; Lee, H.; Kim, G. A surface-modified poly(ε-caprolactone) scaffold comprising variable nanosized surface-roughness using a plasma treatment. *Tissue Eng. Part. C Methods* **2014**, *20*, 951–963. [CrossRef] [PubMed]
42. Qian, Y.; Cheng, Y.; Ouyang, Y.; Yuan, W.E.; Fan, C. Multilayered spraying and gradient dotting of nanodiamond–polycaprolactone guidance channels for restoration of immune homeostasis. *NPG Asia Mater.* **2019**, *11*. [CrossRef]
43. Gill, H.S.; Prausnitz, M.R. Does needle size matter? *J. Diabetes Sci. Technol.* **2007**, *1*, 725–729. [CrossRef] [PubMed]
44. Kalsoom, U.; Peristyy, A.; Nesterenko, P.N.; Paull, B. A 3D printable diamond polymer composite: A novel material for fabrication of low cost thermally conducting devices. *RSC Adv.* **2016**, *6*, 38140–38147. [CrossRef]
45. Vandrovcova, M.; Douglas, T.E.L.; Mróz, W.; Musial, O.; Schaubroeck, D.; Budner, B.; Syroka, R.; Dubruel, P.; Bacakova, L. Pulsed laser deposition of magnesium-doped calcium phosphate coatings on porous polycaprolactone scaffolds produced by rapid prototyping. *Mater. Lett.* **2015**, *148*, 178–183. [CrossRef]
46. Turlybekuly, A.; Pogrebnjak, A.D.; Sukhodub, L.F.; Sukhodub, L.B.; Kistaubayeva, A.S.; Savitskaya, I.S.; Shokatayeva, D.H.; Bondar, O.V.; Shaimardanov, Z.K.; Plotnikov, S.V.; et al. Synthesis, characterization, in vitro biocompatibility and antibacterial properties study of nanocomposite materials based on hydroxyapatite-biphasic ZnO micro- and nanoparticles embedded in Alginate matrix. *Mater. Sci. Eng. C* **2019**, *104*, 109965. [CrossRef]

nanomaterials

MDPI

Article

Reagent-Free Colorimetric Assay for Galactose Using Agarose Gel Entrapping Nanoceria and Galactose Oxidase

Phuong Thy Nguyen, Hee Tae Ahn and Moon Il Kim *

Department of BioNano Technology, Gachon University, 1342 Seongnamdae-ro, Sujeong-gu, Seongnam, Gyeonggi 13120, Korea; nnphuongthy18@gmail.com (P.T.N.); venice4@naver.com (H.T.A.)

* Correspondence: moonil@gachon.ac.kr; Tel.: +82-31-750-8563

Received: 8 April 2020; Accepted: 5 May 2020; Published: 8 May 2020

Abstract: A reagent-free colorimetric method for galactose quantification using a composite of cerium oxide nanoparticles (nanoceria) and galactose oxidase (Gal Ox) entrapped in an agarose gel was developed. In the presence of galactose, the Gal Ox entrapped within the agarose gel catalyzed the oxidation of galactose to generate H_2O_2, which induced a color change from white to intense yellow. This reaction occurred without any chromogenic substrate. This color transition is presumed to be due to the H_2O_2-mediated alteration of the oxidation state of cerium ions present on the surface of the nanoceria. The intensity of color change was quantified by acquiring an image with a conventional smartphone, converting the image to cyan-magenta-yellow-black (CMYK) mode, and subsequently analyzing the image using the ImageJ software. Using this strategy, galactose concentration was specifically determined with excellent sensitivity of as low as 0.05 mM. The analytical utility of the assay was successfully verified by correctly determining diverse levels of galactose in human serum, which is enough to diagnose galactosemia, a genetic disorder characterized by the malfunctioning of enzymes responsible for galactose metabolism. The assay employing a hydrogel composite with entrapped nanoceria and Gal Ox, is a simple, cost-effective, and rapid colorimetric assay for galactose quantification, without using any chromogenic reagent. This cost-effective method has great potential for the diagnosis of galactosemia and is highly promising in comparison to the laborious instrumentation-based methods currently in use.

Keywords: reagent-free colorimetric assay; galactose determination; nanoceria; agarose gel; galactosemia diagnosis

---

1. Introduction

Galactose, one of the breakdown products of lactose (a disaccharide composed of one glucose and one galactose molecule) in milk, is an essential nutrient in the human body. It is also an important marker for diagnosing a disease that affects newborns, galactosemia. Galactosemia is a congenital metabolic disorder caused by the deficiencies of relevant enzymes involved in galactose metabolism [1]. Galactosemia is categorized into types I, II, and III, which occur due to the malfunctioning of galactose-1-phosphate uridyltransferase (GALT), galactokinase (GALK), and uridine diphosphate galactose-4-epimerase (GALE), respectively [2,3]. Among these, type I galactosemia is recognized as the most common and severe form of galactosemia and is termed "classic galactosemia" [4]. Patients with classic galactosemia cannot metabolize galactose and galactose-1-phosphate, leading to their accumulation in blood, and thus causing life-threatening complications, such as jaundice, cataracts, hepatic failure, and kidney dysfunction [5,6]. Thus, screening newborns for galactosemia, by evaluating the levels of galactose and galactose-1-phosphate or the efficiency of relevant enzymes in the blood, is crucial for its early diagnosis and effective treatment.

Many analytical methods have been developed for the diagnosis of galactosemia using dried blood spot specimens of newborns. The most traditional method is based on bacterial growth inhibition, as suggested by Paigen et al. [7]. This assay uses a mutant *Escherichia coli* strain to check its resistance or sensitivity to bacteriophage infection in the presence or absence of galactose. The bacteria are resistant to bacteriophage infection in the presence of galactose, whereas they are sensitive and are infected by bacteriophages in its absence. This method quantifies galactose levels by measuring the bacterial growth zone around the blood spots. Other methods employing genetically-engineered bacteria have also been reported [8]. For example, a transferase-deficient *E. coli* mutant whose growth is inhibited by galactose was employed for galactose quantification and the subsequent diagnosis of galactosemia. A galactose-1-phosphate uridylyltransferase-encoding gene (*galT*) knockout strain of *E. coli* exhibited growth proportional to glucose levels but not galactose levels, while another *E. coli* strain grew normally in the presence of either glucose or galactose. Galactose concentrations were quantified by measuring the growth differences between the two strains. All these microbiological methods can detect galactose with high specificity; however, they generally require long incubation times for bacterial growth. Additionally, microbial strains are sometimes unstable, yielding inaccurate results, leading to repetitive testing to measure galactose levels more accurately and exclude any false positive results. Currently, galactose levels are usually determined using diverse analytical instruments, such as high-performance liquid chromatography (HPLC), gas chromatography/mass spectrometry (GC/MS), and tandem mass spectrometry (MS/MS) [9–11]. Assays employing these techniques enable sensitive and accurate analyses for galactose, but are expensive and often complicated to operate due to many pre-/post-treatment steps; hence, they are not suitable for rapid screening of newborns.

A microscale well-plate assay based on a coupled enzymatic reaction involving alkaline phosphatase and galactose dehydrogenase was more recently developed [12]. In this assay, alkaline phosphatase catalyzes the production of galactose from galactose-1-phosphate in the presence of galactose dehydrogenase and NAD^+ to yield NADH. The resulting NADH activates diaphorase to convert iodonitrotetrazolium violet into a red colored product, formazan, which can be visually detected. Another reported cascade enzyme reaction utilizes a multi-catalyst system composed of Gal Ox and a peroxidase mimicking nanozyme [13]. In the presence of galactose, Gal Ox generates H_2O_2, which subsequently activates peroxidase-mimicking magnetic nanoparticles and converts a colorimetric substrate of 2,2′-azino-bis(3-ethylbenzo-thiazoline-6-sulfonic acid) diammonium salt (ABTS) into a green product. These enzymatic methods are quite sensitive and quantitative, but multiple components, including the colorimetric substrate and complicated procedures, still limit their widespread use. These enzymes have been also widely employed to develop electrochemical assays, which could yield the highest sensitivity with the lowest detection limits without the requirement of pre-/post-treatment steps [14]. Some of them based on the unique screen-printed electrode and Langmuir–Blodgett film-deposited electrode showed impressive detection limits, even down to the low nanomolar level; however, the appropriate electrodes and electrometer should be required for the measurement of galactose level [15,16]. Thus, there is a great incentive to develop a simpler, more rapid, and more economical method for galactose determination.

Recently, we reported the use of a multi-catalyst system consisting of glucose oxidase or cholesterol oxidase in combination with nanoceria as a colorimetric strategy that helps with detecting the corresponding target molecules [17]. This assay involves an H_2O_2-mediated color change of nanoceria without any addition of colorimetric reagents; however, it still depends on spectrophotometry for quantification. This feature critically hinders its widespread applications, particularly in facility-limited environments. As an advancement of this work for practical on-site application, we herein developed a composite system, which consists of nanoceria and Gal Ox as an oxidative enzyme within an agarose gel. Essential analytical features of the system, such as selectivity, sensitivity, and detection precision, can be assessed using images acquired with a smart cellular phone, which is quite suitable for instrumentation-free, point-of-care testing (POCT) environments.

2. Experimental Section

2.1. Materials

Nanoceria (<5 nm particle size), agarose (low-gelling temperature), galactose oxidase from *Dactylium dendroides* (Gal Ox), galactose, arabinose, fructose, lactose, maltose, glucose, sodium acetate, and human serum were purchased from Sigma-Aldrich (Milwaukee, WI, USA). The average particle size of the nanoceria was checked by transmission electron microscopy (TEM) analysis. Hydrogen peroxide (35%) was purchased from Junsei Chemical Co. (Tokyo, Japan). All the chemicals were of analytical grade or higher, and all solutions were prepared with distilled (DI) water purified using a Milli-Q Purification System (Millipore, Billerica, MA, USA).

2.2. Preparation of Agarose Gel Composites Entrapping Nanoceria Only (Agarose_Nanoceria) and Both Gal Ox and Nanoceria (Agarose_Nanoceria + Gal Ox)

Agarose gel composites were synthesized by slightly modifying the reported procedure [17]: 2% (*w*/*v*) agarose solution was first prepared by dissolving agarose in distilled water using microwave treatment for 1–2 min. To entrap nanoceria in an agarose matrix, they were dissolved in sodium acetate buffer (100 mM, pH 5.3) at a concentration of 15 mg/mL and then mixed with 2% (*w*/*v*) agarose solution in a 1:4 volume ratio. After mixing rigorously, 15 mL of the mixture was poured into a petri dish (90 × 15 mm, SPL Lifesciences, Pocheon, Korea) and left to solidify at room temperature (RT) for 1 h. Next, the agarose gel was imprinted with a 200 μL pipet tip as the mold, to make a tablet-type agarose_nanoceria (5.4 mm in diameter), which fitted perfectly into the wells of a 96-well plate. An agarose composite with embedded Gal Ox and nanoceria (agarose_nanoceria + Gal Ox) was prepared by mixing nanoceria (15 mg/mL in sodium acetate buffer (100 mM, pH 5.3), Gal Ox (6 mg/mL in sodium acetate buffer (100 mM, pH 5.3), and 2% agarose solution in a 1:1:3 volume ratio and following the same procedure as that described for agarose_nanoceria preparation.

The agarose_nanoceria or agarose_nanoceria + Gal Ox tablets were washed with deionized water three times and then used in further experiments. The amounts of protein and nanoceria that leached from the agarose composite into the supernatant were measured using the BCA protein assay (PIERCE, Rockford, IL, USA) and inductively coupled plasma atomic emission spectroscopy (ICP-AES) (Polyscan 60E, Thermo Jarrell, Franklin, MA, USA), respectively. The final concentrations of Gal Ox and nanoceria within the agarose composite were calculated using the differences between the initial and the leached concentrations. The prepared agarose composites were also analyzed by field emission TEM (Tecnai, FEI, Amsterdam, The Netherlands) in scanning transmission electron microscopy (STEM) and energy dispersive X-ray spectroscopy (EDS) imaging modes. For scanning electron microscopy (SEM) and Fourier transform infrared (FT-IR) analysis, the samples were freeze-dried for 2 days and then analyzed with a field emission scanning electron microscope (Magellan 400) and an FT-IR Spectrometer (Nicolet iS50, Thermo Fisher Scientific, Madison, WI, USA), respectively. The hydrodynamic diameters of free nanoceria were determined by dynamic light scattering (DLS) analysis using a Zetasizer Nano-ZS (Malvern Co., Worcestershire, UK).

2.3. Determination of H_2O_2 Using Agarose_Nanoceria

H_2O_2 concentration was determined using a transparent 96-well plate as follows. A mixture containing free nanoceria (0.2 mg) or an agarose_nanoceria tablet (one tablet approximately contains 0.2 mg nanoceria) and H_2O_2 (20 μL at varied concentrations) in 180 μL sodium acetate buffer (100 mM, pH 5.3) was incubated at RT for 10 min. The images of the resultant well-plate were then acquired using a smartphone (GALAXY S8 NOTE, Samsung, Seoul, Korea), and were subjected to quantitative image processing via conversion to "CMYK" mode using ImageJ software (version 1.48, NIH, Bethesda, MD, USA).

2.4. Determination of Galactose Using Agarose_Nanoceria + Gal Ox

Galactose concentration was determined using a 96-well plate as follows. A mixture containing agarose_nanoceria + Gal Ox (one tablet approximately containing 0.2 mg nanoceria and 0.08 mg Gal Ox) and galactose (20 μL at diverse concentrations) in 180 μL sodium acetate buffer (100 mM, pH 5.3) was added into the wells of a 96-well plate and incubated at 37 °C for 30 min. The resultant well-plate was directly used to obtain images with a smartphone, and other procedures were the same as those used for H_2O_2 determination.

Long-term stability of the agarose_nanoceria + Gal Ox tablets was examined in aqueous buffer (100 mM sodium acetate, pH 5.3) at RT under static conditions, by measuring color intensity in response to 10 mM galactose at predetermined time points. The relative color intensity (%) was calculated using the ratio of the residual color intensity to the original one. Synthetic reproducibility of the agarose_nanoceria + Gal Ox tablets was also assessed by measuring the color intensities of the tablets toward 10 mM galactose, which were prepared from 5 different synthetic batches.

To determine the amount of galactose in human serum, the original concentration of galactose in serum was first determined using a galactose assay kit (Sigma-Aldrich). Pre-determined amounts of galactose were then added to human serum to get spiked samples that represent normal, boundary, and galactosemia levels of galactose. The concentration of galactose in each spiked sample (20 μL) was measured using the same procedure as mentioned above. The precision and reproducibility of the assays were assessed by determining the recovery rate (recovery (%) = measured value/actual value × 100) and the coefficient of variation (CV (%) = SD/average × 100).

3. Results and Discussion

3.1. Construction of Agarose Composite for Colorimetric Determination of Galactose

A composite system for galactose quantification, containing nanoceria and Gal Ox simultaneously entrapped within an agarose gel matrix, was established (Figure 1). It was constructed by simply mixing nanoceria and Gal Ox in 2% (*w*/*v*) aqueous agarose solution, followed by gelation at RT for 1 h. This process resulted in about 18% and 8% loading for nanoceria and Gal Ox, respectively. We envisioned that the agarose composite (agarose_nanoceria + Gal Ox) would serve as an efficient colorimetric galactose biosensor capable of being used for diagnosing galactosemia, one of the major newborn metabolic disorders. In the presence of galactose, Gal Ox in the composite catalyzed the oxidation of galactose to produce H_2O_2, which subsequently interacted with the nanoceria to induce the oxidation of surface-exposed Ce^{3+} to Ce^{4+} species, and produced peroxide complexes at the nanoceria surface [18,19]. This promoted the vivid color change of nanoceria from white to intense yellow without any chromogenic substrate. Images were acquired using a smartphone under laboratory conditions, and quantitative information was obtained by simple image processing by converting the obtained JPEG-type image to "CMYK" mode using ImageJ software.

To obtain structural insights for the synthesized agarose composites, TEM analyses coupled with EDS elemental mapping were performed (Figure 2). The images revealed that the nanoceria were incorporated within the hydrogel. Elemental mapping images of Ce, N, and O elements, which are the main components of nanoceria, Gal Ox, and agarose, respectively, demonstrated that Gal Ox was homogeneously distributed throughout the agarose gel matrix. Nanoceria seemed to prefer adsorption on the rough and porous gel surfaces to their dispersion; however, there were still many nanoceria observed throughout the gel matrix, which may contribute to an enhanced colorimetric response compared with that of free nanoceria, which are prone to aggregation when suspended in an aqueous solution [20]. SEM images also showed the existence of nanoceria within the agarose gel, and their porous surfaces (Figure 3a,b), which may facilitate the transfer of galactose molecules through the gel matrix. Further TEM and SEM images are also shown in Figure S1. We also performed FT-IR analysis to analyze and compare the chemical structures of agarose_nanoceria with those of nanoceria and agarose (Figure 3c). The peaks observed at 773, 894, and 932 cm^{-1} in the spectrums of both agarose and agarose_nanoceria were attributed to 3,6-anhydro-L-galactose, which is the skeletal structure of the agarose gel [21,22]. The characteristic peak at 480 cm^{-1} for both nanoceria and agarose_nanoceria was attributed to the Ce–O–O stretching vibration, clearly indicating that nanoceria were encapsulated in the agarose gel matrix [23,24].

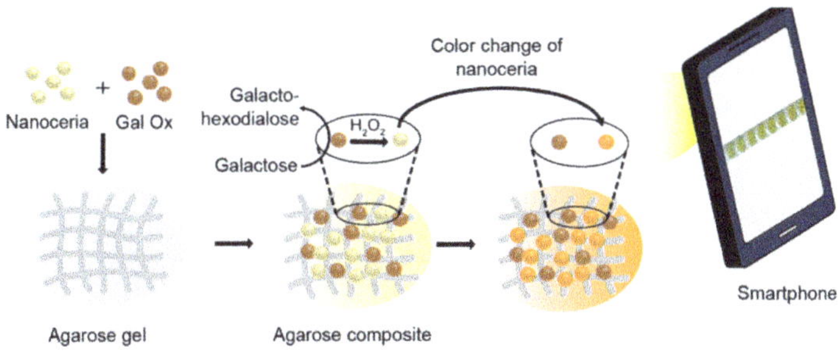

Figure 1. Schematic illustration of the agarose composite consisting of nanoceria and Gal Ox for the smartphone-mediated, reagent-free colorimetric determination of galactose.

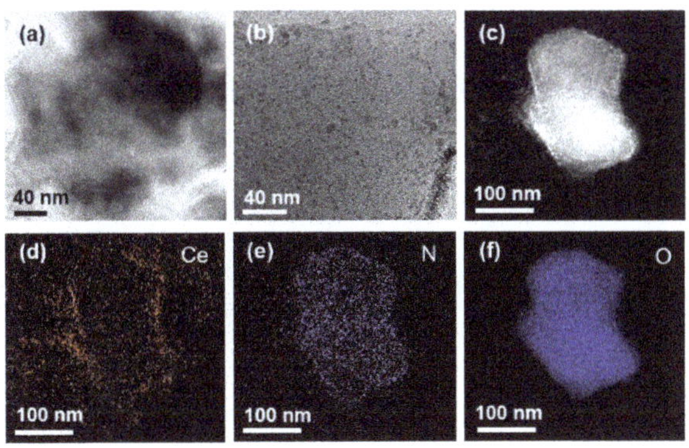

Figure 2. TEM images of (**a**) free agarose gel and (**b**) agarose_nanoceria. (**c**) STEM images of agarose composite entrapping both nanoceria and Gal Ox, and the corresponding EDS maps of (**d**) Ce, (**e**) N, and (**f**) O elements.

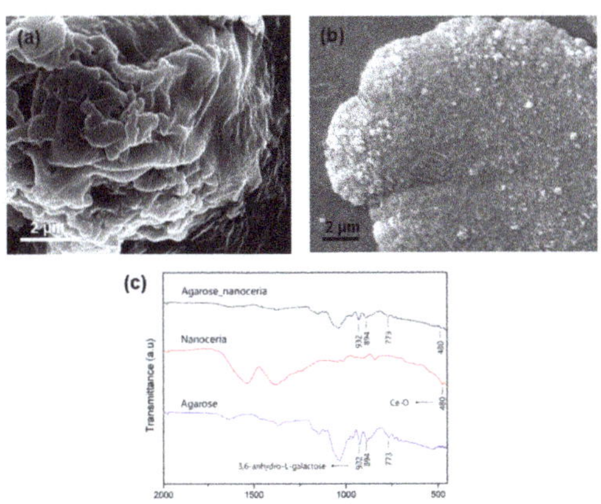

Figure 3. SEM images of (**a**) free agarose gel and (**b**) the agarose_nanoceria mixture. (**c**) FT-IR spectra of agarose_nanoceria, nanoceria, and agarose.

3.2. Vivid Colorimetric Responses of Agarose_Nanoceria toward H_2O_2

The colorimetric responses of agarose_nanoceria were examined in the presence of H_2O_2 and were quantified by capturing images with a smartphone (Figure 4). This assay is based on the principle that nanoceria develop a yellowish color proportional to the H_2O_2 concentration in a sample. Free nanoceria were used as the control. For image processing with the ImageJ software, the real images of agarose_nanoceria and free nanoceria were converted into the CMYK mode and analyzed. The investigations on the effects of buffer pH and composition, gel density, and temperature on the color intensity of agarose_nanoceria, showed that sodium acetate buffer (100 mM, pH 5.3), 2% (*w*/*v*) agarose, and 37 °C were ideal assay conditions (Figure S2). Although the incubation at 37 °C resulted in maximal color intensity, the incubation at RT was adopted rather than 37 °C because of the experimental convenience and the sufficiently acceptable sensing performance of agarose_nanoceria at RT (about 90% of maximum color intensity). Under the optimized conditions, the agarose_nanoceria enabled clear color change from white to intense yellow with 10 mM H_2O_2 during a 10 min reaction. The color change was similarly observed for free nanoceria as well; however, the intensity of the color developed was ~2-fold lower than that of agarose_nanoceria (Figure 4a,b), possibly due to the aggregated nature of free nanoceria [20], which is in contrast to the well-dispersed nature of nanoceria entrapped within the agarose matrix (Figure 2b). DLS analyses also showed that free nanoceria yielded ~300 nm of hydrodynamic diameter, while their single particle size was estimated to be only ~5 nm (Figure S3), confirming their aggregation in our assay conditions. Based on the enhanced color changes of agarose_nanoceria in the presence of H_2O_2, multiple samples containing various concentrations of H_2O_2 were analyzed (Figure 4c). The color intensity increased with increasing H_2O_2 concentrations, having a dynamic linear range of 0.1–1.5 mM (R^2 = 0.9968), and the limit of detection (LOD) was calculated to be 0.017 mM (Figure 4d), which is enough for coupling with Gal Ox and creating an efficient galactose assay system.

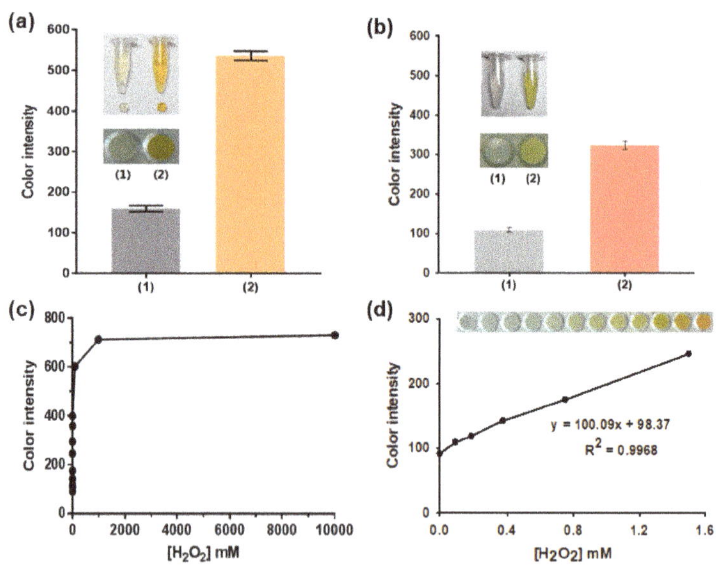

Figure 4. Colorimetric detection of H_2O_2 using agarose_nanoceria composite. Real images of the color-change reaction in the (1) absence and (2) presence of H_2O_2 (10 mM) and the corresponding color intensities observed using (**a**) agarose_nanoceria and (**b**) free nanoceria. (**c**) A dose-response curve for H_2O_2 determination using the agarose_nanoceria, and (**d**) the corresponding linear calibration plot with a real image. Each error bar represents the standard deviation of three independent measurements.

3.3. *Analytical Capabilities of the Agarose Composite: Specificity, Linearity, Sensitivity, and Precision for Galactose Determination*

Through the simple colorimetric reaction of agarose_nanoceria + Gal Ox without any colorimetric reagent, galactose was specifically determined as it showed a vivid color change from white to intense yellow in 30 min (Figure 5a). On the contrary, no significant color change was observed for the negative control samples, such as arabinose, fructose, maltose, and glucose, which confirmed the excellent specificity of the assay system towards the target galactose. As already known, Gal Ox shows significant reactivity toward lactose as substrate; there was a considerable signal detected for lactose in this study. However, this undesirable reaction would not interfere with our galactose determination assay, since lactose is always hydrolyzed into glucose and galactose by lactase in our body, and hence, lactose rarely appears in a considerable amount in human blood [25].

Through the analysis of dose-response curve, the LOD for galactose was determined to be as low as 0.05 mM, with a linear range from 0.1 to 1.5 mM (Figure 5b). The LOD value of the assay system was calculated by this formula: LOD = 3 × δ/slope, where δ is the standard deviation of blank and slope is the slope of calibration curve [26]. Considering that the cut-off value of galactosemia is about 0.44 mM (8 mg/dL), the current system is suitable to distinguish patients with galactosemia from normal persons [27]. Although several fluorescence-based enzymatic assays showed higher sensitivity, the LOD and linear range values of our system are among the best results describing colorimetric detection of galactose (Table S1). Furthermore, our assay system solely enables reagent-free colorimetric determination of galactose, which is quite advantageous in practical applications.

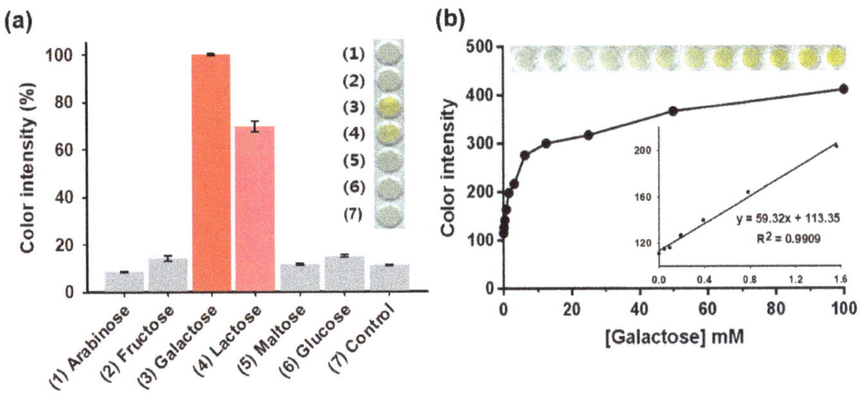

Figure 5. (**a**) Real images of the selective colorimetric detection of galactose using the agarose composite entrapping nanoceria and Gal Ox with the corresponding color intensities. A 5 mM concentration of galactose and lactose was used, while 50 mM of other carbohydrates was used in the experiments. (**b**) A dose-response curve, real images, and the corresponding linear calibration plots for galactose determination using the composite entrapping nanoceria and Gal Ox. Each error bar represents the standard deviation of three independent measurements.

We generally used the agarose composites directly after the synthesis; however, they could be stored at 4 °C for one-month period of time without any significant loss of activity (data not shown). Then, the effects of entrapment of Gal Ox or nanoceria within agarose gel on stability were examined. To this, residual activities of the agarose_nanoceria + Gal Ox were determined along with those of free Gal Ox incubated in the presence of either agarose_nanoceria or free nanoceria. During 10 days of incubation at RT under static conditions, our agarose_nanoceria + Gal Ox almost retained their original activity, while free Gal Ox-mediated samples displayed significant losses of activity (Figure S4), showing the vivid enhancement in stability. Furthermore, agarose_nanoceria +

Gal Ox composites, prepared from different synthetic batches, yielded similar color intensity with low variations, demonstrating the synthetic reproducibility (Figure S5).

Finally, we evaluated the diagnostic capability of the agarose_nanoceria + Gal Ox-based assay using clinical human serum samples containing levels of galactose corresponding to normal, boundary, and galactosemia (normal; ≤0.44 mM, boundary; 0.44–1.11 mM, and galactosemia; >1.11 mM) [28]. The original amount of galactose in the serum samples was first determined using a galactose assay kit, and a pre-determined amount of galactose was then added to establish the representative levels. According to the experimental results, the serum galactose levels were quantitatively determined with excellent precision, yielding CVs ranging from 4.80% to 6.90% and recoveries ranging from 94.47% to 105.86% (Table 1), demonstrating the excellent reproducibility and reliability of this assay. These results prove that the reagent-free colorimetric galactose assay described herein may serve as a promising analytical tool to diagnose galactosemia in clinical settings.

Table 1. Detection precision of the agarose_nanoceria_Gal Ox-based assay system for the determination of galactose levels in spiked human serum samples.

	Original Amount (mM)	Added Galactose Concentration (mM)	Expected Galactose Concentration (mM)	Measured [a] Galactose Concentration (mM)	SD [b]	CV [c] (%)	Recovery [d] (%)
Normal		0.139	0.199	0.188	0.012	6.38	94.47
Boundary	0.060	0.589	0.649	0.687	0.033	4.80	105.86
High		1.333	1.393	1.377	0.095	6.90	98.85

[a] The average value of five successive measurements. [b] The standard deviation (SD) of five successive measurements. [c] Coefficient of variation = (SD/mean) × 100. [d] Recovery = (measured value/expected value) × 100.

4. Conclusions

Herein, we prepared a composite system consisting of nanoceria and Gal Ox simultaneously entrapped within an agarose gel matrix as a reagent-free colorimetric sensor to detect galactose. The developed system yielded excellent selectivity, sensitivity, and linearity by processing the real image acquired on-site using a smartphone. The clinical application of the bioassay was successfully demonstrated by accurately determining the galactose concentrations from clinical human serum samples. Thus, this agarose composite-based galactose assay might serve as a promising analytical tool for a rapid, robust, and convenient method for galactose quantification, which is essential for the diagnosis of galactosemia in newborns. Since the current system enabled rapid visual detection of the target molecule in an instrumentation-free and cost-effective manner without the involvement of any chromogenic substrate, it should find practical applications in POCT environments.

Supplementary Materials: The following figures are available online at http://www.mdpi.com/2079-4991/10/5/895/s1. Figure S1: (a) TEM and (b) SEM images of agarose_nanoceria. Figure S2: Effects of (a) temperature, (b) buffer system (pH and composition), and (c) gel density on the level of color intensity for the detection of H_2O_2. Figure S3: Hydrodynamic size distributions of free nanoceria measured by DLS. Figure S4: Stability of agarose_nanoceria + Gal Ox and free Gal Ox in the presence of either agarose_nanoceria or free nanoceria. Figure S5: Color intensity toward 10 mM galactose using agarose_nanoceria + Gal Ox composites prepared from five different batches. Table S1: An overview on the essential assay attributes of optical determination of galactose.

Author Contributions: P.T.N., conceptualization, investigation, and writing—original draft preparation. H.T.A., conceptualization and investigation. M.I.K., conceptualization, supervision, and writing—review and editing. All authors have read and agreed to the published version of the manuscript.

Funding: This work was supported by the Center for Research and Development of the Police Science and Technology grant funded by the Korea National Police Agency (KNPA) (PA-K000001-2019-401) and the National Research Foundation of Korea (NRF) grant funded by the Korea government (Ministry of Science and ICT (NRF-2019R1A2C1087459)).

Conflicts of Interest: The authors declare no conflict of interest.

References

1. Weese, S.J.; Gosnell, K.; West, P.; Gropper, S.S. Galactose content of baby food meats: Considerations for infants with galactosemia. *J. Am. Diet. Assoc.* **2003**, *103*, 373–375. [PubMed]
2. Holton, J. Galactose disorders: An overview. *J. Inherit. Metab. Dis.* **1990**, *13*, 476–486. [CrossRef] [PubMed]
3. Holden, H.M.; Rayment, I.; Thoden, J.B. Structure and function of enzymes of the Leloir pathway for galactose metabolism. *J. Biol. Chem.* **2003**, *278*, 43885–43888. [CrossRef] [PubMed]
4. Jeong, J.-S.; Kwon, H.-J.; Yoon, H.-R.; Lee, Y.-M.; Choi, T.-Y.; Hong, S.-P. A pulsed amperometric detection method of galactose 1-phosphate for galactosemia diagnosis. *Anal. Biochem.* **2008**, *376*, 200–205. [CrossRef]
5. Bosch, A.M. Classical galactosaemia revisited. *J. Inherit. Metab. Dis.* **2006**, *29*, 516–525. [CrossRef]
6. Bennett, M.J. Galactosemia diagnosis gets an upgrade. *Clin. Chem.* **2010**, *56*, 690–692. [CrossRef]
7. Paigen, K.; Pacholec, F.; Levy, H.L. A new method of screening for inherited disorders of galactose metabolism. *J. Lab. Clin. Med.* **1982**, *99*, 895–907.
8. Woo, M.-A.; Kim, M.I.; Cho, D.; Park, H.G. Cell-based galactosemia diagnosis system based on a galactose assay using a bioluminescent *Escherichia coli* array. *Anal. Chem.* **2013**, *85*, 11083–11089. [CrossRef]
9. Hu, O.Y.-P.; Hu, T.-M.; Tang, H.-S. Determination of galactose in human blood by high-performance liquid chromatography: Comparison with an enzymatic method and application to the pharmacokinetic study of galactose in patients with liver dysfunction. *J. Pharm. Sci.* **1995**, *84*, 231–235.
10. Nishimura, Y.; Tajima, G.; Bahagia, A.D.; Sakamoto, A.; Ono, H.; Sakura, N.; Naito, K.; Hamakawa, M.; Yoshii, C.; Kubota, M. Differential diagnosis of neonatal mild hypergalactosaemia detected by mass screening: Clinical significance of portal vein imaging. *J. Inherit. Metab. Dis.* **2004**, *27*, 11–18. [CrossRef]
11. Ning, C.; Segal, S. Plasma galactose and galactitol concentration in patients with galactose-1-phosphate uridyltransferase deficiency galactosemia: Determination by gas chromatography/mass spectrometry. *Metabolism* **2000**, *49*, 1460–1466. [CrossRef] [PubMed]
12. Diepenbrock, F.; Heckler, R.; Schickling, H.; Engelhard, T.; Bock, D.; Sander, J. Colorimetric determination of galactose and galactose-1-phosphate from dried blood. *Clin. Biochem.* **1992**, *25*, 37–39. [CrossRef]
13. Kim, M.I.; Shim, J.; Li, T.; Woo, M.-A.; Cho, D.; Lee, J.; Park, H.G. Colorimetric quantification of galactose using a nanostructured multi-catalyst system entrapping galactose oxidase and magnetic nanoparticles as peroxidase mimetics. *Analyst* **2012**, *137*, 1137–1143. [CrossRef] [PubMed]
14. Kanyong, P.; Krampa, F.D.; Aniweh, Y.; Awandare, G.A. Enzyme-based amperometric galactose biosensors: A review. *Microchim. Acta* **2017**, *184*, 3663–3671. [CrossRef]
15. Kanyong, P.; Pemberton, R.M.; Jackson, S.K.; Hart, J.P. Amperometric screen-printed galactose biosensor for cell toxicity applications. *Anal. Lett.* **2016**, *49*, 236–244. [CrossRef]
16. Sharma, S.K.; Singhal, R.; Malhotra, B.D.; Sehgal, N.; Kumar, A. Langmuir-Blodgett film based biosensors for estimation of galactose in milk. *Electrochim. Acta* **2004**, *49*, 2479–2485. [CrossRef]
17. Kim, D.H.; Hur, J.; Park, H.G.; Kim, M.I. Reagentless colorimetric biosensing platform based on nanoceria within an agarose gel matrix. *Biosens. Bioelectron.* **2017**, *93*, 226–233. [CrossRef]
18. Heckert, E.G.; Karakoti, A.S.; Seal, S.; Self, W.T. The role of cerium redox state in the SOD mimetic activity of nanoceria. *Biomaterials* **2008**, *29*, 2705–2709. [CrossRef]
19. Scholes, F.H.; Soste, C.; Hughes, A.E.; Hardin, S.G.; Curtis, P.R. The role of hydrogen peroxide in the deposition of cerium-based conversion coatings. *Appl. Surf. Sci.* **2006**, *253*, 1770–1780. [CrossRef]
20. Feng, X.; Sayle, D.C.; Wang, Z.L.; Paras, M.S.; Santora, B.; Sutorik, A.C.; Sayle, T.X.; Yang, Y.; Ding, Y.; Wang, X.; et al. Converting ceria polyhedral nanoparticles into single-crystal nanospheres. *Science* **2006**, *312*, 1504–1508. [CrossRef]
21. Prasad, K.; Mehta, G.; Meena, R.; Siddhanta, A.K. Hydrogel-forming agar-graft-PVP and κ-carrageenan-graft-PVP blends: Rapid synthesis and characterization. *J. Appl. Polym. Sci.* **2006**, *102*, 3654–3663. [CrossRef]
22. Trivedi, T.J.; Rao, K.S.; Kumar, A. Facile preparation of agarose–chitosan hybrid materials and nanocomposite ionogels using an ionic liquid via dissolution, regeneration and sol–gel transition. *Green Chem.* **2014**, *16*, 320–330. [CrossRef]
23. Das, J.; Han, J.W.; Choi, Y.-J.; Song, H.; Cho, S.-G.; Park, C.; Seo, H.G.; Kim, J.-H. Cationic lipid-nanoceria hybrids, a novel nonviral vector-mediated gene delivery into mammalian cells: Investigation of the cellular uptake mechanism. *Sci. Rep.* **2016**, *6*, 1–13. [CrossRef]

24. McDevitt, N.T.; Baun, W.L. Infrared absorption study of metal oxides in the low frequency region (700-240 cm^{-1}). *Spectrochim. Acta* **1964**, *20*, 799–808. [CrossRef]
25. Ford, J.D.; Haworth, J.C. The estimation of galactose in plasma using galactose oxidase. *Clin. Chem.* **1964**, *10*, 1002–1006. [CrossRef] [PubMed]
26. Kim, H.; Lee, J.U.; Song, S.; Kim, S.; Sim, S.J. A shape-code nanoplasmonic biosensor for multiplex detection of Alzheimer's disease biomarkers. *Biosens. Bioelectron.* **2018**, *101*, 96–102. [CrossRef] [PubMed]
27. Park, I.S.; Cho, H.J.; Lee, D.H.; Song, J.H. Galactosemia detected by meonatal screening test. *J. Korean Pediatr. Soc.* **2003**, *46*, 440–446.
28. Jensen, U.G.; Brandt, N.J.; Christensen, E.; Skovby, F.; Nørgaard-Pedersen, B.; Simonsen, H. Neonatal screening for galactosemia by quantitative analysis of hexose monophosphates using tandem mass spectrometry: A retrospective study. *Clin. Chem.* **2001**, *47*, 1364–1372. [CrossRef] [PubMed]

nanomaterials

MDPI

Article

Fabrication of Stable Nanofiber Matrices for Tissue Engineering via Electrospinning of Bare Laser-Synthesized Au Nanoparticles in Solutions of High Molecular Weight Chitosan

Viraj P. Nirwan [1,2], Ahmed Al-Kattan [2,*], Amir Fahmi [1,*] and Andrei V. Kabashin [2,3]

1 Faculty of Technology and Bionics, Rhine-Waal University of Applied Science, Marie-Curie-Straβe 1, 47533 Kleve, Germany

2 Aix Marseille University, CNRS, LP3 (UMR 7341), 13288 Marseille, France

3 MEPhI, Institute of Engineering Physics for Biomedicine (PhysBio), 115409 Moscow, Russia

* Correspondence: ahmed.al-kattan@univ-amu.fr (A.A.-K.); Amir.Fahmi@hochschule-rhein-waal.de (A.F.); Tel.: +33-(0)4918-292-86 (A.A.-K.); +49-2821-806-73 (A.F.)

Received: 14 June 2019; Accepted: 19 July 2019; Published: 24 July 2019

Abstract: We report a methodology for the fabrication of neutralized chitosan-based nanofiber matrices decorated with bare Au nanoparticles, which demonstrate stable characteristics even after prolonged contact with a biological environment. The methodology consists of electrospinning of a mixture of bare (ligand-free) laser-synthesized Au nanoparticles (AuNPs) and solutions of chitosan/polyethylene oxide (ratio 1/3) containing chitosan of a relatively high molecular weight (200 kDa) and concentration of 3% (*w*/*v*). Our studies reveal a continuous morphology of hybrid nanofibers with the mean fiber diameter of 189 nm ± 86 nm, which demonstrate a high thermal stability. Finally, we describe a protocol for the neutralization of nanofibers, which enabled us to achieve their structural stability in phosphate-buffered saline (PBS) for more than six months, as confirmed by microscopy and FTIR measurements. The formed hybrid nanofibers exhibit unique physicochemical properties essential for the development of future tissue engineering platforms.

Keywords: electrospinning; laser ablation; nanofibers; chitosan; poly(ethylene oxide) (PEO); Au nanoparticles; neutralization

1. Introduction

The elaboration of artificial scaffold platforms, capable of replacing and/or repairing failing tissues or full organs, is still challenging for the scientific community [1–4]. Most efforts are devoted to the fabrication of nanostructured scaffolds, which could mimic mesoporous morphologies of a real extracellular matrix (ECM) and offer additional therapy and diagnostics (theranostics) modalities [5–7]. Similar to conventional techniques, such as solvent casting and particle leaching, electrospinning is now extensively explored to elaborate structured biocompatible and biodegradable nanofibrous scaffolds. Such a technique can offer a series of advantages over traditional methods, including (i) the possibility of working with a variety of materials, including synthetic and natural polymers and their composites and (ii) the capability of generating micro- to nano-scale nanofibers having sophisticated special 3D designs [8]. Moreover, electrospun nanofibers can be employed as matrices for the encapsulation of drugs, as well as the incorporation of biological materials (e.g., proteins, DNA, etc.) and nanofunctional elements [9–13].

Chitosan is one of the most extensively exploited naturally derived biopolymers, which is widely used to develop novel electrospun nanofibrous matrices for a variety of biomedical applications, including anti-microbial, antifungal, wound healing, drug, and gene delivery [14–18]. Chitosan is characterized by a good biocompatibility and biodegradability, while the presence of NH_2 groups on a chitosan nanofiber surface make it suitable for the immobilization of enzymes and negatively charged proteins. Moreover, a highly reactive chitosan surface can offer additional opportunities for its functionalization with nano-engineered particles to improve physicochemical characteristics (e.g., electrical and mechanical properties) and enable biological (e.g., antibacterial) or other theranostic functionalities [5–19]. However, the spinnability of chitosan polymer is relatively poor due to the presence of hydrogen bonds between polysaccharide chains, leading to its high crystallinity and weak solubility in most solvents. Therefore, acid solutions, such as acetic acid, trifluoroacetic acid (TFA), and ionic liquids are typically employed to facilitate the dissolution of chitosan. Moreover, due to a relative shortness of the chitosan chain, a polymer-stretching effect during electrospinning process is ineffective, which results in the formation of discontinued jets. To ensure the jet continuity and uniformity, one has to electrospin chitosan together with poly(ethylene oxide) (PEO) polymer [20–22]. Here, low molecular weight chitosan (102 kDa) looks preferable to ensure its complete dissolution, but such a choice typically requires high concentration of PEO to avoid the formation of small chitosan chains in the solution: This leads to a low concentration of chitosan in the matrix. On the other hand, the use of chitosan having too high a molecular weight (>310 kDa) results in the formation of gel structures, which are not suitable for electrospinning [23]. The employment of chitosan having reasonably high molecular weight (~200 kDa) looks a good compromise, but such a choice requires the elaboration and optimization of the whole electrospinning procedure [24].

Various chitosan/PEO formulations have been tested at different ratios of chitosan/PEO to obtain homogenous electrospun nanofibers with different characteristics [18,23–26]. In particular, we recently elaborated a novel procedure to fabricate chitosan/PEO nanofibers containing a relatively high molecular weight chitosan (200 kDa) and we decorated them with laser-synthesized Si and Au nanoparticles (SiNPs and AuNPs) as functional additives [5,19]. This procedure was based on the co-spinning of chitosan/PEO formulations (1% *w*/*v* of chitosan, ratio 1:4), together with nanoparticles prepared by methods of laser ablation, as described in our previous studies [27–36]. The uniqueness of laser-synthesized nanoparticles consists in their bare (ligand-free) and uncontaminated surface, which can provide a series of advantages for applications, including cancer diagnoses and therapies under external stimuli [32–35], biofuel cells [31], and Surface Enhanced Raman Spectrometry (SERS) [36], etc. As we showed in References [5,19], the incorporation of functional additives, such as Au and Si nanoparticles, into the chitosan/PEO matrix can provide a series of advantages, including: (i) A decrease of fiber size, which promises the improvement of its surface reactivity; (ii) an improvement of thermal stability at a high temperature; and (iii) the possibility for enabling additional theranostic modalities based on unique properties of laser-synthesized nanomaterials [5]. However, the total concentration of chitosan in the structure of nanofibrous remained relatively low compared to the co-spinning agent (ratio 1:4), which reduced the efficiency of the chitosan matrix and could lead to an unacceptably high rate of dissolution of chitosan/PEO nanofibers in an aqueous environment due to the presence of NH_3^+ [23,37,38].

In this article, we report a further advancement of the electrospinning procedure in order to improve the properties of the formed AuNPs-decorated hybrid chitosan/PEO nanofibers. Here, we show the possibility for increasing concentration of 200 kDa chitosan up to 3% (*w*/*v*) using 10% (*v*/*v*) acetic acid with a chitosan/PEO ratio of 1:3 in the electrospinning solution, which renders possible the solution of the dissolubility problem. Moreover, we describe a simple protocol of neutralization of NH_3^+ in the alkaline/alcoholic solution of 5 M NaOH or 1 M K_2CO_3 to ensure the structural stability of nanofiber membranes in biological media. Novel formulations of chitosan/PEO functionalized with bare laser-synthesized AuNPs look very promising for tissue engineering applications.

2. Materials and Methods

2.1. Materials

A medium molecular weight chitosan (200 kDa) and a PEO powder with the molecular weight of 300 kDa were purchased from Sigma-Aldrich (Darmstadt, Germany). A target (1 cm × 1 cm) of high purity Au (99.999%) was purchased from GoodFellow (Cambridge, United kingdom) to fabricate bare AuNPs. Chitosan was solubilized using acetic acid (99–100%) purchased from SCS GmbH (Sigmarszell, Germany). For the neutralization stage, methanol (99.9%), ethanol (99.5%), and K_2CO_3 were purchased from Carl Roth (Karlsruhe, Germany), while NaOH was purchased from VWR International GmbH (Langenfeld, Germany).

2.2. Methods

2.2.1. Laser-Ablative Synthesis of Bare AuNPs

Bare laser-synthesized AuNPs were prepared using methods of femtosecond ablation and fragmentation in deionized water [27,30,31,33]. Briefly, a gold target (99.99%, GoodFellow, Cambridge, United Kingdom) was placed at the bottom of the glass vessel filled with deionized water (18.2 MΩ cm). A 2.3 mm diameter beam from a Yb:KGW laser (Amplitude Systems, 1025 nm, 480 fs, 1 kHz) was focused with a 75 mm lens on the surface of a target. The target was moved constantly in the focusing plane with a speed of 0.5 mm/s, while keeping the same thickness of the liquid (1 cm) above the target. The concentration of AuNPs was determined by the calculation of the weight loss of the target during the ablation process.

2.2.2. Preparation of Electrospinning Solutions

A total of 2 mL of medium molecular weight chitosan was dissolved at a concentration of 3% (*w*/*v*) using 10% (*v*/*v*) acetic acid. In a separate cuvette, we prepared 4 mL of PEO with a concentration of 8% (*w*/*v*). The solutions were then mixed in the ratio of 1:3, respectively. A concentrated AuNPs solution was added to the polymer solution and ultra-sonicated for 2 h, which ensured the removal of bubbles and a homogenous dispersion of nanoparticles. Concentration of bare laser ablated AuNPs was measured to be 0.15 g L^{-1}. The nanoparticle solution was concentrated before the addition to the electrospinning solution by heating it at 30 °C in an oven for several hours. Such a procedure led to 3-decreasing of the solution volume. A total of 2 mL of the concentrated solution 0.45 g L^{-1} was used to functionalize the electrospinning polymer solutions. This sample was denoted as Ch-AuNPs. A second sample of nanofibers without AuNPs (Ch-0) was prepared as a reference for comparative studies. In this case, 2 mL of AuNP solutions were replaced by 2 mL of PEO. The procedure was carried out in such a way that the final concentration of the polymer was the same as in the main sample (1:3 final ratio by volume). All steps of the preparation protocol are presented in Table 1.

Table 1. Description of preparation steps for main and reference samples.

Main Sample Ch-AuNPs	Reference Sample Ch-0
3% (*w*/*v*) chitosan in 10% (*v*/*v*) acetic acid	3% (*w*/*v*) chitosan in 10% (*v*/*v*) acetic acid
8% (*w*/*v*) PEO in 2 mL conc. AuNPs in deionized water + 2mL deionized water	8% (*w*/*v*) PEO in 4 mL deionized water
1:3 final ratio by volume	1:3 final ratio by volume.

2.2.3. Electrospinning Hybrid Multifunctional Nanofibers

The fabrication of nanofibers was done using specialized equipment purchased from IME medical electrospinning technologies. A climate-controlled setup was employed to ensure constant environmental conditions and fixed parameters during the process. An electrospinning setup contained a rotating cylinder set at 2300 rpm. Electrospinning solutions were transferred into a 5 mL syringe, which was connected to a programmable pump providing the constant feed rate of 0.3 mL h^{-1}. The pump was connected via a Leur-Lock and PTFE (Polytetrafluoroethylene) tube with a blunt 0.8 mm-diameter needle at one end of a spinneret, where 14 kV was applied to the solution. The distance between the collector and spinneret was kept at 15 cm based on previous optimization studies. The collector had a negative voltage of −1 kV and was wrapped in aluminium foil to collect the fibers. Climate conditions were fixed for all experiments at 25 °C temperature and 50% of humidity.

2.3. Morphological and Chemical Analysis

A high-resolution transmission electron microscopy (HR-TEM) system (JEOL JEM 3010, Croissy Sur Seine, France) was used to characterize laser-synthesized AuNPs. Thanks to ImageJ software (National Institutes of Health (NIH), Bethesda, MA, USA), statistic measurements were performed on more than 200 AuNPs to determine diameter size distribution. A scanning electron microscope was used as a primary analysis method to verify if the electrospinning process led to the formation of nanofibers. Due to the limited charging effect on samples, it was unnecessary to proceed to the sputter coating step. Size distributions and morphology of fibers were characterized using scanning electron microscopy (SEM), and a subsequent treatment of images using ImageJ® software. JSM-IT100 InTouchScope™ system (Tokyo, Japan) operating at 5–20 kV accelerating voltages was the main microscope used at lower magnifications. A DSM 982 Gemini Zeiss system (Marly le Roi, France) operating at accelerating voltage of 20 kV was used to produce higher magnification micrographs. Both SEM systems were coupled to energy dispersive X-ray analysis. Chemical characteristics were evaluated by Fourier transform infrared (FTIR) spectroscopy (Perkin–Elmer, Cambridge, MA, USA) equipped with a universal attenuated total reflection (ATR) sampling accessory. The measurement histogram was performed by analyzing 2 micrographs; one from center of the collector, where most of the fiber deposition takes place, and another from the edge of the collector. The frequency distribution, as obtained using integrating ImageJ results with the origin, ranged from 100 to 15,000.

2.3.1. Thermal Analysis

Thermal properties of nanofibers were studied with the help of thermogravimetric analysis (TGA, Perkin–Elmer TGA 4000 system, Billerica, MA, USA) and differential scanning calorimetry (DSC, Perkin–Elmer DSC 8000 system, Billerica, MA, USA). A total of 7 mg and 5 mg of nanofibers were used for each TGA and DSC measurement, respectively. A TGA heating chamber was flushed with nitrogen at 20 mL min^{-1}, preventing the combustion of the samples, and heated at the rate of 10 °C min^{-1}. Similarly, DSC was carried out under a nitrogen environment with a constant flow rate of 21 mL min^{-1} and heating rate was kept at 5 °C min^{-1}. The operation temperature was between 30 and 700 °C during TGA studies and between 30 and 180 °C during DSC analysis.

2.3.2. Neutralization of Chitosan/PEO Nanofibers

A total of two protocols were investigated based on two solutions of 1M K_2CO_3, dissolved in 15 mL of ethanol (70%), and 5M NaOH dissolved in 15 mL of methanol (70%). The nanofibers were immersed in the solutions for a total of 20 h at ambient conditions. The immersion process was repeated 3 times by changing solutions after every immersion step to ensure effective neutralization. The samples were then washed with distilled water and dried at room temperature for 24 h before SEM-EDX and FTIR analyses to assess their structural stability and chemical composition.

3. Results and Discussion

As mentioned above, the electrospinning of pure chitosan is quite challenging due to its poor solubility and short chain length. In most cases, 1% (*w*/*v*) concentration of chitosan is used at variable chitosan/PEO ratios. Recently, chitosan/PEO nanofibers functionalized with bare laser synthesized NPs were successfully electrospun [19]. In view to increase the available bioactivity of functional chitosan, new formulations have been tested. Based on our previous study, where we successfully fabricated chitosan/PEO nanofibers at a chitosan/PEO ratio of 1:4, the concentration of chitosan was increased from 1% (*w*/*v*) to 3% (*w*/*v*) and the chitosan/PEO ratio was increased from 1:4 to 1:3. Moreover, the concentration of acetic acid used to dissolve chitosan was also decreased down to 10% (*v*/*v*), as compared to highly acidic 90% (*v*/*v*) solutions used in our previous study [5,19]. Punctual structural analyses performed, thanks to SEM measurements, revealed that the nanofibers are homogenous without observable beads with a mean diameter of 189 nm ± 100 nm (Figure 1a,b).

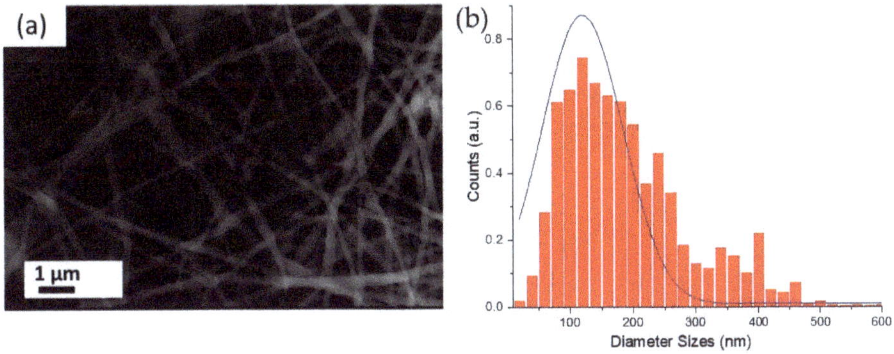

Figure 1. SEM image of hybrid chitosan/poly(ethylene oxide) (PEO) nanofibers (**a**) and corresponding size distribution for nanofiber thicknesses (**b**).

Such initial composition of chitosan/PEO was thus used to prepare a novel solution containing highly concentrated AuNPs (0.09 mg mL^{-1} in electrospinning solution). A typical HR-TEM image of AuNPs formed by laser ablation and corresponding AuNPs size distribution is given in Figure 2.

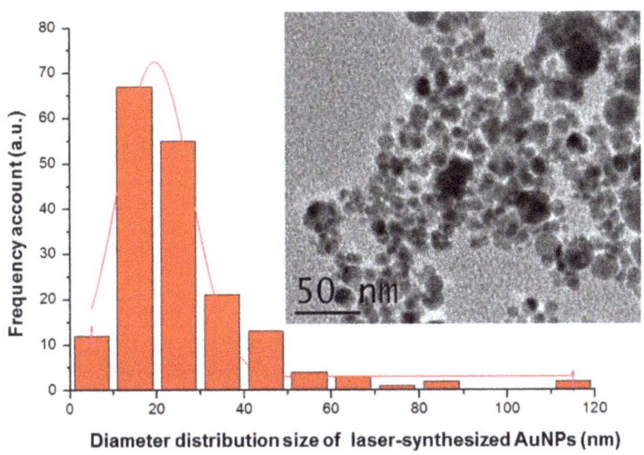

Figure 2. Typical high-resolution transmission electron microscopy (HR-TEM) image (inset image) of laser-synthesized AuNPs and corresponding nanoparticle size distribution.

SEM-EDX examinations of chitosan/PEO nanofibers functionalized with AuNPs revealed a homogenous network of cylindrical nanofibers without any beads. All of these results are consistent with conclusions of our previous study [5], which evidenced similar morphologies of fibers electrospun with and without bare AuNPs. In addition, the fibers exhibited a mean diameter of 189 nm ± 86 nm, as measured on micrographs from different sample regions, using ImageJ® software and fitted with a Gaussian approximation (blue curve) (Figure 3). AuNPs were clearly resolvable on the fiber surface, while the fiber matrix was still homogeneous. Energy-dispersive X-ray spectroscopy (EDX) confirmed the metallic nature of the bare laser AuNPs. The observed affinity between the nanofibers and the AuNPs was likely due to electrostatic interactions between the polycationic fiber surface with positively charged NH_3^+ groups and negatively charged (−23.1 ± 2.61 mV) Au nanoparticles [27,30–33]. It is worth noting that we recorded a 3.5-fold decrease of fiber diameter in comparison with chitosan/PEO nanofiber formulations studied in our previous work [5,19]. It is important that this result was obtained with 1:3 ratio of chitosan/PEO, promising higher reactivity of the nanofiber surface.

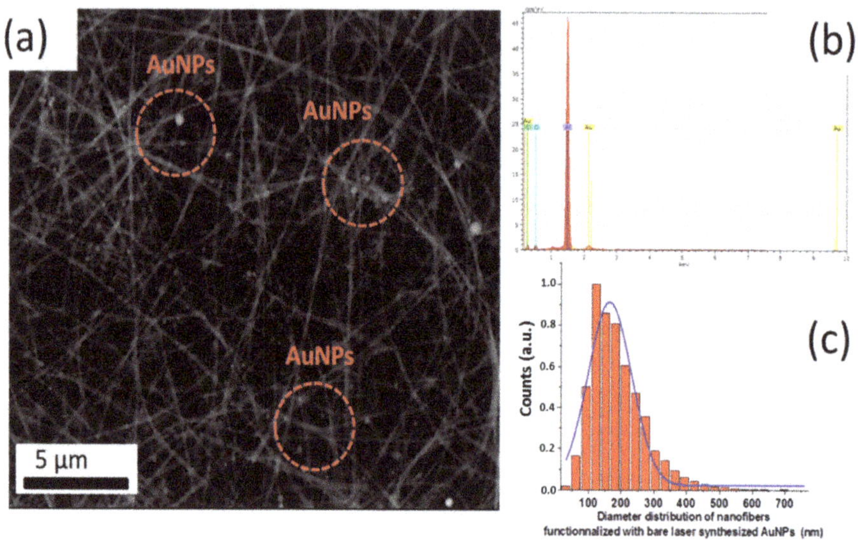

Figure 3. SEM image of hybrid chitosan/PEO nanofibers functionalized with bare laser-synthesized AuNPs (**a**); corresponding energy-dispersive X-ray spectrometry (EDX) spectrum (**b**); and size distribution of functionalized hybrid chitosan/PEO-AuNPs complex (**c**).

In our tests, we used TGA and DSC analyses to examine thermal properties of formed nanofibers and the impact of AuNPs on them. A TGA curve (Figure 4a) revealed that both Ch-0 and Ch-AuNPs exhibit the same behavior degradation. However, it is interesting to note that almost 10 weight % of the reference (Ch-0) is remaining after the heating cycle while 5 weight % of the functionalized nanofibers (Ch-AuNPs) was left after the program ended. This is likely due to the incorporation of AuNPs that act as heating spots leading to the maximum degradation of the polymer. The derivative curve of the TGA analysis was also provided in Figure 4b. When heated beyond their melting point, the nanofibers demonstrated a loss of weight. The weight loss region could be observed for both nanofibers functionalized with AuNPs (Ch-AuNPs) and no-functionalized nanofibers denoted as the reference sample (Ch-0) after 250 °C, which could be interpreted as the initialization of mass loss and could be attributed to the breaking of tertiary bonds in the polymer structure. The major weight loss region started at 347 °C and 366 °C for Ch-0 and Ch-AuNPs samples, respectively. The width of the peak evidenced a narrow range of temperatures, during which the maximum loss/degradation took place. Additionally, the peak temperature for functionalized nanofibers shifted up compared to the reference samples (from 418 to 423 °C).

Subsequently, DSC was carried out to record phase transition changes, which could take place e.g., at nanofiber melting points after their functionalization with gold nanoparticles. Here, both types of nanofibers displayed classical DSC curves. As shown in Figure 5, an endothermic peak could be observed for the reference nanofibers (Ch-0) at 59.10 °C, corresponding to the melting transition of PEO, while for Ch-AuNPs nanofibers, a similar peak was observed at 57.86 °C. Additionally, the normalized peak heat required to enable those transitions was 5.78 mW/mg and 6.81 mW/mg for Ch-0 and Ch-AuNPs, respectively. The calculated enthalpy for these samples provided 19.21 J/g and 16.96 J/g for Ch-AuNPs and Ch-0 samples, as measured by the analysis of the area under the curves. The nanoparticles used for functionalization absorbed additional heat and, therefore, more heat was required to enable a similar transition in functionalized nanofibers. Additionally, a second event was observed after the melting transition of nanofibers for both Ch-AuNPs and Ch-0. At 147.52 °C an exothermic peak could be observed for the reference sample and was at 134.90 °C for the functionalized nanofibers, which could be attributed to the Tg of the chitosan present in both samples. Therefore, it could be presumed that the heat provided while doing the analysis was not only limited by the nanofibers, but also by absorbing nanoparticles. Both TGA and DSC showed that the functionalization of nanofibers using bare laser AuNPs provided extra thermal stability in terms of absorbed heat and degradation rate. It should be noted that TGA and DSC data were highly reproducible based on the analysis of many samples.

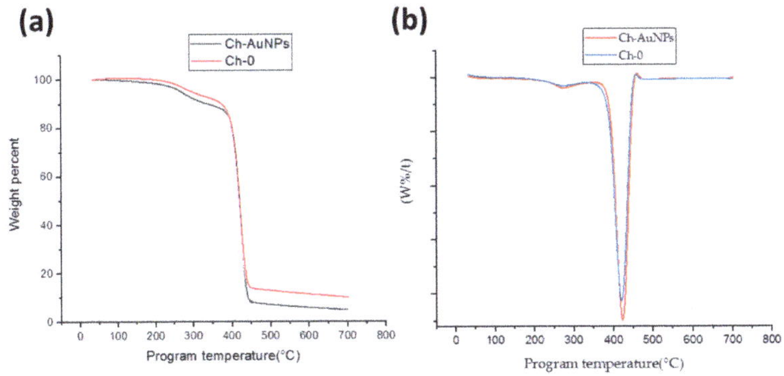

Figure 4. Thermogravimetric analysis (TGA) thermograms (**a**) with corresponding derivative thermogravimetric DTG curves (**b**) of chitosan/PEO nanofibers functionalized with bare laser AuNPs.

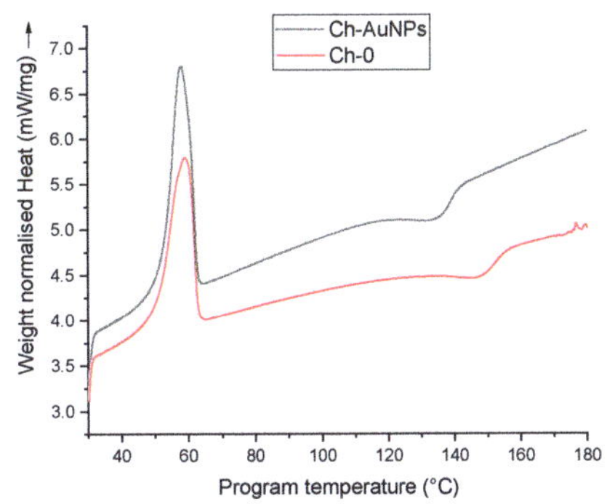

Figure 5. Differential scanning calorimetry (DSC) thermogram curve of chitosan/PEO reference (Ch-0) nanofibers and nanofibers functionalized with bare laser AuNPs (Ch-AuNPs).

The chemical composition of nanofibers was analyzed using FTIR before and after the functionalization. As shown in Figure 6, characteristic peaks could be observed for both constituent polymers (chitosan and PEO). The functionalization of nanofibers with bare laser AuNPs did not change chemical composition. Both the reference and the functionalized nanofibers displayed a broad band of NH_2 and OH, stretching from 3500–3000 cm^{-1}. A peak for amide absorption could also be observed at 1570 cm^{-1} for both functionalized and reference nanofibers [39–41]. A sharp and strong absorption peak attributed to CH_2 stretching in PEO could also be observed at 2885 cm^{-1}. The spectra beyond 1500 seems to be dominated with intense PEO bands at 1145, 1095, and 1059 cm^{-1}, complimentary to C-O-C stretching vibrations [41,42]. The functionalization of nanofibers using bare laser AuNPs seems to have no impact on the chemical signature of nanofibers, as observed by FTIR

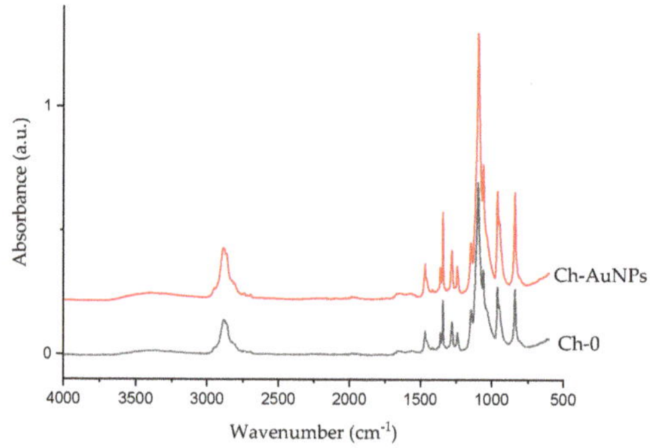

Figure 6. FTIR chitosan/PEO nanofibers functionalized with bare laser AuNPs.

As mentioned in the introduction section, stability of chitosan in aqueous media is a crucial issue. High reactivity of the NH_3^+ group can lead to a rapid dissolution of chitosan/PEO complexes, making them hardly useful for applications in biological systems [28]. Hence, an additional neutralization step is needed to ensure the mechanical stability of the fibrous structure and its non-dissolvability in biological media. As mentioned earlier, two methods were tested to transform the NH_3^+ group into the NH_2 insoluble form: (i) 1M K_2CO_3 dissolved in ethanol (70%) and (ii) 5M NaOH in methanol (70%). After the treatment, the samples were washed in distilled water and immersed in phosphate-buffered saline (PBS) for 24 h to test the neutralization efficacy. For the first method employing 1M K_2CO_3, the overall structure of the treated nanofibers had a non-woven mat-like morphology (Figure 7a).

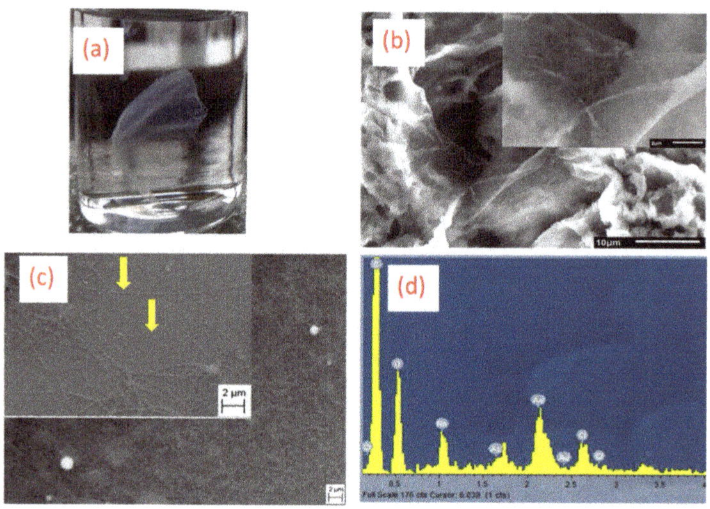

Figure 7. Macroscopic photo of chitosan/PEO functionalized with bare laser AuNPs neutralized with 1M of K_2CO_3 immersed in phosphate-buffered saline (PBS) for 24 h (**a**) with corresponding SEM images (**b**) (The inner image is given as illustration at higher magnification). SEM images of functionalized chitosan/PEO nanofibers with bare laser AuNPs neutralized with 5M NaOH (indicated by yellow arrows) in two different areas (**c**) with the corresponding EDX spectrum (**d**).

However, SEM examination of samples after their immersion in PBS solution for 24 h revealed the loss of the original fibrous structure, as one could not resolve individual nanofibers (Figure 7b), suggesting ineffective neutralization of NH_3^+. In contrast, when 5M NaOH solution in 70% methanol was used for neutralization, the overall nanofibers morphology was conserved at nano and macro scales, as it was indicted by the yellow arrows (Figure 7c). SEM analyses clearly showed that most of the fibers were present, while the presence of AuNPs could still be resolved by EDX analysis (Figure 7d). Moreover, we found the possibility for conserving the structural integrity of nanofibers over six months in PBS after neutralization with 5M NaOH. As shown in a macroscopic photo of nanofibers (Figure 8a,b), the immersion of nanofiber structures in biological PBS fluid did not lead to the dissolution of nanofibers or a dramatic change of their morphology. It is visible that some swelling and shrinkage effects could take place, but these structural changes do not look critical for projected applications, such as biomaterial matrices.

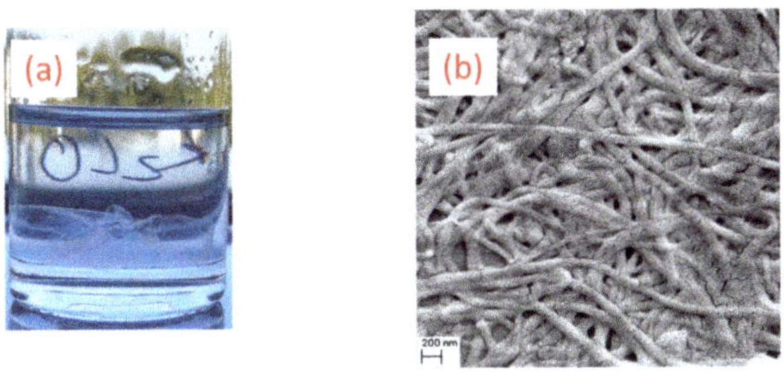

Figure 8. Macroscopic photo of functionalized chitosan/PEO with bare laser AuNPs immersed in PBS over six months (**a**) with the corresponding SEM image (**b**).

To confirm the conservation of chemical composition of chitosan/PEO nanofibers after their treatment by 1M of K_2CO_3 and 5M of NaOH, additional FTIR was carried out (Figure 9). Our analysis revealed that a large part of the PEO was removed, which was especially visible when 5M of NaOH was applied. In fact, main bands of pure PEO including 1467 cm^{-1}, 1277 cm^{-1}, 1343 cm^{-1}, and 1283 cm^{-1} were not resolvable after such a treatment by NaOH, while a characteristic peak of chitosan amine at 3500 cm^{-1} became dominant. A secondary amine peak at 1680 cm^{-1} was also resolvable after both treatments, although this peak is often overshadowed by peaks from PEO [5,41]. Based on the first morphological assessment and FTIR studies, we can conclude that the neutralization of secondary amine in chitosan/PEO with 5M NaOH was successful. However, to make decisive conclusions on the efficiency of neutralization of AuNPs-decorated chitosan/PEO nanofibers and their potential for applications in biological systems, additional parameters (DSC, FTIR, TGA, size distribution, morphology) should be examined in detail. Results of such studies will be published elsewhere.

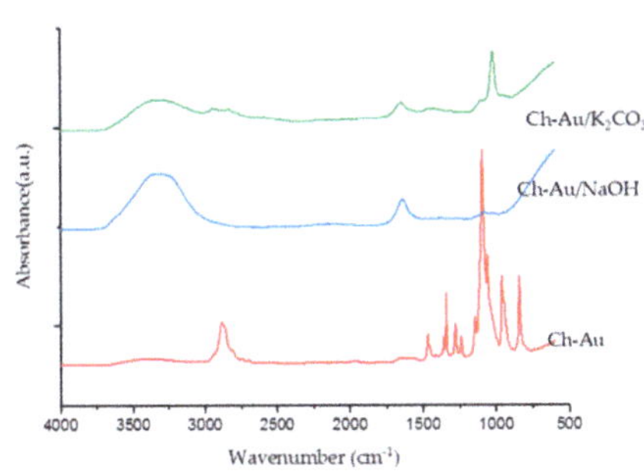

Figure 9. FTIR spectra of chitosan/PEO nanofibers functionalized with AuNPs before neutralization (Ch-Au) and after neutralization with K_2CO_3 (Ch-Au/K_2CO_3) or NaOH (Ch-Au/NaOH).

4. Conclusions

In summary, we prepared novel formulations of hybrid chitosan/PEO nanofibers, with the ratio of 1:3 at a relatively high concentration of chitosan 3% (*w*/*v*), and functionalized these nanofibers with bare laser-synthesized AuNPs. Thanks to systematic physicochemical analysis, improvements on the structural characteristics of the fibers were clearly demonstrated compared to previously reported nanofiber matrices [5,19]. First, microscopic measurements revealed a homogenous network of cylindrical fibers with AuNPs dispersed homogenously on the nanofibers surface without any effect on the chemical composition of the nanofibers, as shown by FTIR. Second, statistical measurements of the fiber size revealed a 3.5-fold decrease (189 ± 86 nm) fitted to a Gaussian monodispersion curve in comparison with the formulations of chitosan reported in our previous study (632 ± 169 nm) [5,19]. Thermal analysis showed that the incorporation of AuNPs into the chitosan/PEO nanofiber formulations improved their thermal stability. Finally, a simple neutralization protocol was proposed, which made possible the conservation of the structural integrity of nanofibers for a period of over six months. This work is a new encouraging step on the development of functional scaffold platforms based on hybrid chitosan/PEO nanofibers functionalized with bare nanoparticles.

Author Contributions: V.P.N. and A.A.-K. conceived the work and performed experiments. All authors analysed the data. V.P.N., A.A.-K., and A.V.K. wrote the manuscript. V.P.N., A.A.-K., A.F., and A.V.K. reviewed the article and added comments and corrections in the final version of the manuscript.

Funding: This research received no external funding.

Acknowledgments: The authors express thanks to the EHYBIOMED project. A.A.-K. acknowledges the German academic exchange service (DAAD - ID 57381327), the International Associated Laboratory (LIA)MINOS project, and the Rhine–Waal University of Applied Science for providing a stipend and laboratories for the PhD program. The authors acknowledge contribution of the Russian Science Foundation (Project 19-72-30012) for the fabrication of several samples of highly calibrated Au nanoparticles. A.V.K. acknowledges support from the MEPhI Academic Excellence Project (Contract No. 02.a03.21.0005).

Conflicts of Interest: The authors declare no conflict of interest.

References

1. Kai, D.; Jin, G.; Prabhakaran, M.P.; Ramakrishna, S. Electrospun synthetic and natural nanofibers for regenerative medicine and stem cells. *Biotechnol. J.* **2013**, *8*, 59–72. [CrossRef] [PubMed]
2. Sridhar, R.; Sundarrajan, S.; Venugopal, J.R.; Ravichandran, R.; Ramakrishna, S. Electrospun inorganic and polymer composite nanofibers for biomedical applications. *J. Biomater. Sci. Polym. Ed.* **2013**, *24*, 365–385. [CrossRef] [PubMed]
3. Stumpf, T.R.; Yang, X.; Zhang, J.; Cao, X. In situ and ex situ modifications of bacterial cellulose for applications in tissue engineering. *Mater. Sci. Eng. C* **2018**, *82*, 372–383. [CrossRef] [PubMed]
4. Patel, H.; Bonde, M.; Srinivasan, G. Biodegradable polymer scaffold for tissue engineering. *Trends Biomater. Artif. Organs* **2011**, *12*, 689–693.
5. Al-Kattan, A.; Nirwan, V.P.; Munnier, E.; Chourpa, I.; Fahmi, A.; Kabashin, A.V. Toward multifunctional hybrid platforms for tissue engineering based on chitosan(PEO) nanofibers functionalized by bare laser-synthesized Au and Si nanoparticles. *RSC Adv.* **2017**, *7*, 31759–31766. [CrossRef]
6. Wu, J.; Xie, L.; Lin, W.Z.Y.; Chen, Q. Biomimetic nanofibrous scaffolds for neural tissue engineering and drug development. *Drug Discov. Today* **2017**, *22*, 1357–1384. [CrossRef] [PubMed]
7. Pierini, F.; Nakielski, P.; Urbanek, O.; Massimiliano, L.; Sio, L.D.; Kowalewski, T.A. Polymer-Based Nanomaterials for Photothermal Therapy: From Light-Responsive to Multifunctional Nanoplatforms for Synergistically Combined Technologies. *Biomacromolecules* **2018**, *11–19*, 4147–4167. [CrossRef] [PubMed]
8. Huang, Z.M.; Zhang, Y.Z.; Kotaki, M.; Ramakrishna, S. A review on polymer nanofibers by electrospinning and their applications in nanocomposites. *Compos. Sci. Technol.* **2003**, *63*, 2223–2253. [CrossRef]
9. Sachlos, E.; Czernuszka, J.T.; Gogolewski, S.; Dalby, M. Making tissue engineering scaffolds work. Review on the application ofsolid freeform fabrication technology to the production of tissue engineeringscaffolds. *Eur. Cells Mater.* **2003**, *5*, 29–40. [CrossRef]

10. Al-Kattan, A.; Nirwan, V.P.; Popov, A.A.; Ryabchikov, Y.V.; Tselikov, G.; Sentis, M.; Fahmi, A.; Kabashin, A.V. Recent advances in laser-ablative synthesis of bare Au and Si NPs and assessment of their prospects for tissue engineering applications. *Int. J. Mol. Sci.* **2018**, *19*, 1563. [CrossRef]
11. Abidian, M.R.; Martin, D.C. Multifunctional nanobiomaterials for neural interfaces. *Adv. Funct. Mater.* **2009**, *19*, 573–585. [CrossRef]
12. Agarwal, S.; Greiner, A.; Wendorff, J.H. Functional materials by electrospinning of polymers. *Prog. Polym. Sci.* **2013**, *38*, 963–991. [CrossRef]
13. Klein, K.L.; Melechko, A.V.; McKnight, T.E.; Retterer, S.T.; Rack, P.D.; Fowlkes, J.D.; Joy, D.C.; Simpson, M.L. Surface characterization and functionalization of carbon nanofibers. *J. Appl. Phys.* **2008**, *103*, 0631301. [CrossRef]
14. Ahmed, S.; Ikram, S. Chitosan Based Scaffolds and Their Applications in Wound Healing. *Achiev. Life Sci.* **2016**, *10*, 27–37. [CrossRef]
15. Zhao, L.M.; Shi, L.E.; Zhang, Z.L.; Chen, J.M.; Shi, D.D.; Yang, J.; Tang, Z.X. Preparation and application of chitosan nanoparticles and nanofibers. *Braz. J. Chem. Eng.* **2011**, *28*, 353–362. [CrossRef]
16. Muxika, A.; Etxabide, A.; Uranga, J.; Guerrero, P.; de la Caba, K. Chitosan as a bioactive polymer: Processing, properties and applications. *Int. J. Biol. Macromol.* **2017**, *105*, 1358–1368. [CrossRef] [PubMed]
17. Dash, M.; Chiellini, F.; Ottenbrite, R.M.; Chiellini, E. Chitosan–A versatile semi-synthetic polymer in biomedical applications. *Prog. Polym. Sci.* **2011**, *36*, 981–1014. [CrossRef]
18. Elsabee, M.Z.; Naguib, H.F.; Morsi, R.E. Chitosan based nanofibers, review. *Mater. Sci. Eng. C* **2012**, *32*, 1711–1762. [CrossRef]
19. Al-kattan, A.; Nirwan, V.P.; Munnier, E.; Chourpa, I.; Popov, A.A.; Al-Kattan, A.; Nirwan, V.P.; Munnier, E.; Tselikov, G.; Ryabchikov, Y.V.; et al. Bare laser-synthesized Si nanoparticles as functional elements for chitosan nanofiber-based tissue. Proc. SPIE. *Synth. Photonics Nanoscale Mater. XV* **2018**, *9*, 10521.
20. Jayakumar, R.; Menon, D.; Manzoor, K.; Nair, S.V.; Tamura, H. Biomedical applications of chitin and chitosan based nanomaterials-A short review. *Carbohydr. Polym.* **2010**, *82*, 227–232. [CrossRef]
21. Torres-Giner, S.; Ocio, M.J.; Lagaron, J.M. Development of active antimicrobial fiber based chitosan polysaccharide nanostructures using electrospinning. *Eng. Life Sci.* **2008**, *8*, 303–314. [CrossRef]
22. Adrienne, H.; Nanofibers, E.C. Electrospinning chitosan-based nanofibers for biomedical application. In *Degree of Master of Science*; Graduate Faculty of North Carolina State University: Raleigh, NC, USA, 2006.
23. Lemma, S.M.; Bossard, F.; Rinaudo, M. Preparation of pure and stable chitosan nanofibers by electrospinning in the presence of poly(ethylene oxide). *Int. J. Mol. Sci.* **2016**, *17*, 2710–2714.
24. Wendorff, J.H.; Agarwal, S.; Greiner, A. *Electrospinning: Materials, Processing, and Applications*; John Wiley & Sons: Hoboken, NJ, USA, 2012; ISBN 9783527320806.
25. Ohkawa, K.; Cha, D.; Kim, H.; Nishida, A.; Yamamoto, H. Electrospinning of Chitosan. *Macromol. Rapid Commun.* **2004**, *25*, 1600–1605. [CrossRef]
26. Vasconcelos, C.L.; Bezerril, P.M.; dos Santos, D.E.S.; Dantas, T.N.C.; Pereira, M.R.; Fonseca, J.L.C. Effect of molecular weight and ionic strength on the formation of polyelectrolyte complexes based on poly(methacrylic acid) and chitosan. *Biomacromolecules* **2006**, *7*, 1245–1252. [CrossRef] [PubMed]
27. Kabashin, V.K.; Meunier, M. Synthesis of Colloidal Nanoparticles during Femtosecond Laser Ablation of Gold in Water. *J. Appl. Phys.* **2003**, *94*, 7941. [CrossRef]
28. Kabashin, A.V.; Meunier, M. Laser-Induced Treatment of Silicon in Air and Formation of Si/SiOx Photoluminescent Nanostructured Layers. *Mater. Sci. Eng. B* **2003**, *101*, 60–64. [CrossRef]
29. Gongalsky, M.B.; Osminkina, L.A.; Pereira, A.; Manankov, A.A.; Fedorenko, A.A.; Vasiliev, A.N.; Solovyev, V.V.; Kudryavtsev, A.A.; Sentis, M.; Kabashin, A.V.; et al. Laser-Synthesized Oxide-Passivated Bright Si Quantum Dots for Bioimaging. *Sci. Rep.* **2016**, *6*, 24732. [CrossRef] [PubMed]
30. Maximova, K.; Aristov, A.; Sentis, M.; Kabashin, A.V. Size-Controllable Synthesis of Bare Gold Nanoparticles by Femtosecond Laser Fragmentation in Water. *Nanotechnology* **2015**, *26*, 65601. [CrossRef] [PubMed]
31. Hebié, S.; Holade, Y.; Maximova, K.; Sentis, M.; Delaporte, P.; Kokoh, K.B.; Napporn, T.W.; Kabashin, A.V. Advanced Electrocatalysts on the Basis of Bare Au Nanomaterials for Biofuel Cell Applications. *ACS Catal.* **2015**, *5*, 6489–6496. [CrossRef]
32. Al-Kattan, A.; Ryabchikov, Y.V.; Baati, T.; Chirvony, V.; Sánchez-Royo, J.F.; Sentis, M.; Braguer, D.; Timoshenko, V.Y.; Estève, M.-A.; Kabashin, A.V. Ultrapure Laser-Synthesized Si Nanoparticles with Variable Oxidation States for Biomedical Applications. *J. Mater. Chem. B* **2016**, *4*, 7852–7858. [CrossRef]

33. Correard, F.; Maximova, K.; Estève, M.-A.; Villard, C.; Roy, M.; Al-Kattan, A.; Sentis, M.; Gingras, M.; Kabashin, A.V.; Braguer, D. Gold Nanoparticles Prepared by Laser Ablation in Aqueous Biocompatible Solutions: Assessment of Safety and Biological Identity for Nanomedicine Applications. *Int. J. Nanomed.* **2014**, *9*, 5415–5430.
34. Baati, T.; Al-kattan, A.; Esteve, M.; Njim, L.; Ryabchikov, Y.; Chaspoul, F.; Hammami, M.; Sentis, M.; Kabashin, A.V.; Braguer, D. Ultrapure Laser-Synthesized Si-Based Nanomaterials for Biomedical Applications: In Vivo Assessment of Safety and Biodistribution. *Sci. Rep.* **2016**, *6*, 1–13. [CrossRef] [PubMed]
35. Kabashin, A.V.; Timoshenko, V.Y. What Theranostic Applications Could Ultrapure Laser-Synthesized Si Nanoparticles Have in Cancer? *Nanomedicine* **2016**, *11*, 2247–2250. [CrossRef] [PubMed]
36. Kögler, M.; Ryabchikov, Y.V.; Uusitalo, S.; Popov, A.; Popov, A.; Tselikov, G.; Välimaa, A.-L.; Al-Kattan, A.; Hiltunen, J.; Laitinen, R.; et al. Bare laser-synthesized Au-based nanoparticles as non-disturbing SERS probes for Bacteria Detection. *J. Biophotonics* **2018**, *11*, e201700225. [CrossRef] [PubMed]
37. Gu, B.K.; Park, S.J.; Kim, M.S.; Kang, C.M.; Kim, J.I.; Kim, C.H. Fabrication of sonicated chitosan nanofiber mat with enlarged porosity for use as hemostatic materials. *Carbohydr. Polym.* **2013**, *97*, 65–73. [CrossRef]
38. Sangsanoh, P.; Supaphol, P. Stability improvement of electrospun chitosan nanofibrous membranes in neutral or weak basic aqueous solutions. *Biomacromolecules* **2006**, *7*, 2710–2714. [CrossRef]
39. Pawlak, A.; Mucha, M. Thermogravimetric and FTIR studies of chitosan blends. *Proc. Thermochim. Acta* **2003**, *396*, 153–166. [CrossRef]
40. Osman, Z.; Arof, A.K. FTIR studies of chitosan acetate based polymer electrolytes. *Electrochim. Acta* **2003**, *48*, 993–999. [CrossRef]
41. Zivanovic, S.; Li, J.; Davidson, P.M.; Kit, K. Physical, mechanical, and antibacterial properties of chitosan/PEO blend films. *Biomacromolecules* **2007**, *8*, 1505–1510. [CrossRef]
42. Pucić, I.; Jurkin, T. FTIR assessment of poly(ethylene oxide) irradiated in solid state, melt and aqeuous solution. *Radiat. Phys. Chem.* **2012**, *81*, 1426–1429. [CrossRef]

nanomaterials

MDPI

Article

Lipidic Liquid Crystalline Cubic Phases and Magnetocubosomes as Methotrexate Carriers

Monika Mierzwa, Adrianna Cytryniak, Paweł Krysiński and Renata Bilewicz *

Faculty of Chemistry, University of Warsaw, Pasteura 1, PL 02-093 Warsaw, Poland; monikaszlezak89@wp.pl (M.M.); a.cytryniak@gmail.com (A.C.); pakrys@chem.uw.edu.pl (P.K.)

* Correspondence: bilewicz@chem.uw.edu.pl

Received: 21 March 2019; Accepted: 16 April 2019; Published: 19 April 2019

Abstract: The release profiles of methotrexate, an anticancer drug, from the monoolein liquid crystalline cubic phases were studied. The cubic phases were used either in the form of a lipidic film deposited onto a glassy carbon electrode surface or in the dispersed form of magnetocubosomes, which are considered a prospective hybrid drug delivery system. Commonly, cubosomes or liposomes are employed, but not in the case of toxic methotrexate, known to block the receptors responsible for folate transport into the cells. The release profiles of the drug from the lipidic films were monitored electrochemically and described using the Higuchi model. They were also modified via changes in temperature; the release was faster, although it deviated from the model when the temperature was increased. Cubic phase nanoparticles (magnetocubosomes) containing hydrophobic magnetic nanoparticles placed in an alternating magnetic field of low frequency and amplitude, stimulated drug release from the suspension, which was monitored spectroscopically. These new biocompatible hybrid nanomaterials in the dispersed form allow to control the release of the drug at the appropriate sites, can be easily separated or relocated under external magnetic field and await further investigations of their in vitro cytotoxicity and in vivo biodistribution.

Keywords: methotrexate; cubic phase; magnetocubosomes; monoolein; liquid crystalline phase; drug delivery system; alternating magnetic field

---

1. Introduction

Methotrexate (MTX), also known as amethopterin, is an antimetabolite of folic acid. Its molecular structure differs from that of the folic acid (FA) in two ways. First, MTX has a four-amino group instead of a hydroxyl group on the pteridine ring, and second, it has a methyl group at the N^{10} position (Scheme S1).

MTX blocks folate synthesis by inhibiting enzyme dihydrofolate reductase, which synthesizes the active form of folate required for DNA and protein synthesis, leading to a major decrease in cancer cell proliferation [1]. It is widely used for the treatment of acute leukemia, malignant lymphoma, osteogenic sarcoma, choriocarcinoma, and other cancers (breast, head, neck, pulmonary, and epidermoid). MTX is also used in the treatment of rheumatoid arthritis or other autoimmune diseases [2,3]. MTX is currently one of the most studied antitumor chemotherapeutic drugs since its therapeutic utility is limited by its high toxicity in the healthy cells of patients whose levels of folates are also seriously reduced [4]. The principal side effects of MTX treatment are myelosuppression, thrombocytopenia, anemia, gastrointestinal mucositis, hepatitis, nausea, fatigue, headache, and/or dyspnea [3–6].

In light of the systemic side effects of anticancer drugs, the design of drug carriers is very important. There are several methods for limiting MTX toxicity through its administration by means of carriers or targeting carriers, e.g., liposomes [7,8], silica materials [9,10], lipid nanoparticles [11,12], chitosan nanoparticles [13], polymeric micelles with magnetic nanoparticles [14], nanorods [15], implants [16], and nanogels [17,18].

A promising carrier is the liquid-crystalline lipidic cubic phase (LCP), which possesses a variety of desirable properties making it a good candidate for drug delivery systems (DDS) [19]. The advantages of nonlamellar LCP nanoparticles such as cubosomes over planar structures and liposomes are connected with the larger membrane surface area to nanoparticle volume and increased drug-loading capacity thanks to the huge internal lipid-water interface. The curvature of the bilayer can be tuned independently of the nanoparticle size, channel size, and the thickness of the lipid layer can be controlled by the selection of lipids used for the formation of the cubic phase and application of dopants [20,21]. An outer block co-polymer stabilizes and makes targeting the carrier possible. Cubosomes retain the internal structure of parent bulk phase, however, in the dispersed system the viscosity of the bulk phase is reduced, which makes it easier to apply.

LCP composed of monoolein (MO) is non-toxic, bioadhesive, and biodegradable. The biodegradability of LCP is due to the fact that monoolein undergoes slow lipolysis in contact with the esterase found in the tissues. The highly structured inverse bicontinuous LCP is composed of two identical, nonintersecting aqueous channels that are surrounded by curved lipid bilayers and can accommodate large amounts of active molecules of any polarity [22]. LCPs are stable in the presence of water, which is an important property in light of their perspective applications for DDS. The rate of drug release from liquid crystals can be modified by either changing the symmetry of the liquid crystal, the dimension of the water channel, or the pH of the environment. Generally, this can be achieved through the manipulation of the lipid composition [23–26]. Previously, we studied the physical and chemical properties of hybrid materials composed of LCPs bearing hydrophilic and hydrophobic magnetic nanoparticles as a material for drug delivery where the outcome was promising [27]. Magnetic nanoparticles and ions are currently of great interest in research related to medical applications [28] and are used as contrast agents for theranostics in medical imaging applications [29]. Few reported studies have examined the effect of the addition of nanoparticles to the liquid crystalline hexagonal phase [30], liposomes [31,32], and cubosomes [33–35], creating a magnetic responsive system. Magnetic nanoparticles with hydrophobic organic shell, localized within the bilayer lipid membrane of liposomes were reported to mechanically rupture these structures in low-frequency alternating magnetic field (LF-AMF) of 50 Hz and 10 mT field strength. No magnetic hyperthermia was involved in such field. When the vesicles were loaded with doxorubicin—A potent cytostatic drug, the relatively mild magnetic stimulus triggered the release of the drug [32]. Montis et al. studied in situ the diffusion of two fluorescent tags, the hydrophobic Octadecyl Rhodamine B and the hydrophilic Rhodamine 110, from the magnetocubosomes using Fluorescence Correlation Spectroscopy during the application of LF-AMF. The application of LF-AMF accelerated the release of the hydrophilic molecules from the water channels of magnetocubosomes [33].

It is known that the use of external alternating magnetic field may adversely affect tissues, including normal tissue toxicity. This effect can be considered in terms of heat production by the interaction of AMF with tissue by its direct heating, induction heating of a material embedded within the tissue (here—magnetic nanoparticles) or both. The second type of heat production is localized within the direct vicinity of superparamagnetic iron oxide nanoparticles and is considered for, e.g., cancer treatment, via the classical hyperthermia response. For the hyperthermia-related treatment the AMF frequencies of kHz to MHz are used, with magnetic field amplitude of ca. 40 to 100 mT [36]. In our studies low-frequency AMF (house frequency) of 50 Hz and 10 mT amplitude is used, and in fact, hyperthermia effect can be neglected as discussed below in Section 3.3. On the other hand, interaction of AMF with tissue can produce heat directly, even though this "material" is of very low conductivity as compared with typical conductive material, like metals. The mechanism that dominates this nonspecific heating results from the production of eddy currents (Foucault's currents) producing heat that scales as: $SAR_{EC} \sim f^2 H^2 r^2$, in which SAR_{EC} is the tissue-specific absorption rate [W/g of tissue], r is the radius of affected region, f and H are the AMF frequency and amplitude, respectively. It was reported [36] that no adverse effects were observed for BALB/c athymic nude mice for AMF amplitudes below 70 mT even at continuous power application (100% duty) for up to 20 min. Generally, it is accepted that high-amplitude AMF (<130 mT) and a frequency of ca. 150 kHz was well tolerated, provided fast heat dissipation. Again, we stress that in the studies reported in this work we have used 50 Hz (3000 times lower frequency) and 10 mT field (4-fold smaller), therefore, it will not induce nonspecific tissue heating.

In this study, we focus on the delivery of MTX loaded monoolein hybrid systems. The MTX load and release studies were conducted for the monoolein cubic phase and cubosomes with and without hydrophobic magnetic nanoparticles present in the lipidic part of the carrier. By taking advantage of the presence of hydrophobic magnetic nanoparticles localized in the lipidic part of magnetocubosomes, we used an alternating magnetic field (AMF) with low frequency to stimulate the release of the drug from the carriers.

2. Experimental

2.1. Materials

All chemicals were of the highest quality and commercially available. Monoolein (1-oleoyl-rac-glycerol) (MO) (>99% pure) was purchased from Hampton Research (Aliso Viejo, CA, USA) and used as received. Methotrexate (N-[4{[2,4-diamino-6-pteridinyl-methyl]-methylamino}benzoyl] glutamic acid) (MTX) was obtained from Fluka (>98% pure, Munich, Germany) and used without further purification. Pluronic F-127 was purchased from Sigma-Aldrich (St. Louis, MO, USA). A 0.2 M phosphate buffer was prepared by titrating 0.2 M di-sodium hydrogen phosphate (>99% pure, ChemPur, Karlsruhe, Germany) with 0.2 M sodium di-hydrogen phosphate (>99% pure, Polish Chemicals Co., Poland) to pH 7.4. All solutions were prepared using Milli Q water (18.2 $M\Omega \cdot cm^{-1}$; Millipore, Bedford, MA, USA). To synthesize the hydrophobic nanoparticles, $FeCl_3 \bullet 6H_2O$ (≥98% pure, Sigma-Aldrich), $FeCl_2 \bullet 4H_2O$ (≥99% pure, Sigma-Aldrich), NH_4OH (25% aqueous solution in H_2O, Chempur), and oleic acid (>99% pure, Sigma-Aldrich) were used.

2.2. Preparation of Cubic Phases and Cubosomes

The LCPs were prepared by adding the appropriate amount of MTX solution to a 0.1 M phosphate buffer in a glass vial with molten MO to obtain 59/1/40, 59.5/0.5/40, or 59.75/0.25/40% (w/w) for the phase MO/MTX/buffer.

Magnetic nanoparticles (MNPs) used to prepare hybrid cubic phases were synthesized as described in our previous report [27]. Briefly, we used the co-precipitation method with oleic acid as the capping agent. The synthesis was carried out under vigorous stirring and an argon atmosphere. As the first step, $FeCl_2 \bullet 4H_2O$ (0.023 mol) and $FeCl_3 \bullet 6H_2O$ (0.046 mol) were dissolved in 150 mL of deionized water. The mixture was heated to 45 °C and then 11 mL of 25% ammonia was added, which caused the appearance of a black precipitate of nanoparticles. After half an hour, 3 mL of oleic acid was added. Then, the suspension was heated to 80 °C and kept at this temperature for one hour. Afterwards, the Fe_3O_4 nanoparticles coated with oleic acid were washed several times with deionized water and ethanol under magnet-assisted sedimentation to remove the excess oleic acid. After thorough washing, the nanoparticles of ca. 7 ± 2 nm in diameter were transferred to hexane as a stable suspension of controlled density. The size distribution and mass contribution of the organic shell (ca. 35%) were discussed in our previous work [27,32]. The magnetic characterization of dry powder nanoparticles was carried out with a QD vibrating sample magnetometer VSM, working at 300 K and stabilized to 0.01 K accuracy, over the magnetic field range from 2.0 to +2.0 T. The saturation magnetization of the as synthesized, oleic acid-coated nanoparticles was ca. 43 emu/g, which was comparable with the literature and our previous work [32].

To prepare the hybrid LCPs, the appropriate amount of MNP dispersion in hexane was added to the molten MO and sonicated. The mixture was left in a desiccator to evaporate the solvent. Then, MTX, in a phosphate buffer solution was added to the mixture. The ratio of the components of the hybrid system MO/NPs/MTX/buffer was 57.5/2/0.5/40% (w/w). The samples were left in tightly sealed vials at room temperature in the dark until they became homogenous and transparent.

The procedure of preparing the magnetocubosome dispersion started with mixing molten MO and MNP dispersion in hexane (34.5 mg of MNPs per 1 g of MO). After the evaporation of the solvent, MTX was added (8.7 mg MTX per 1 g of MO). The solution was then mixed and followed by sonication until it became homogeneous. Then, the 1% F-127 solution was added to obtain 94% w/w and the mixture was dispersed by ultrasonication using a Sonics Vibra cell for 20 min (3 s "on", 5 s "off") at an amplitude of 40% of maximum output (130 watt, 30 kHz).

2.2.1. Dynamic Light Scattering (DLS) and Zeta Potential

The hydrodynamic diameter of the cubosomes was determined using a Malvern Zetasizer instrument (Nano ZS, Malvern, UK) fitted with a 4 mW He–Ne laser (λ = 632.8 nm) as the light source with a scattering angle of 173°. Samples were equilibrated for 2 min before measurement. The zeta potential was measured using the same instrument. The values were reported as averages from 3 measurements of each sample.

2.2.2. Cryogenic Transmission Electron Microscopy (Cryo-TEM)

A droplet of the cubosome sample was deposited on a lacey copper grid. The sample was then blotted with filter paper inside a climate chamber (Vitrobot Mark IV, FEI, Hillsboro, OR, USA). The temperature was kept at 22 °C and 100% humidity. Afterwards, the grid was plunged into a liquid ethane bath cooled with liquid nitrogen. The cooled gird was maintained at a temperature of −170 °C using a cryo holder 626 (Gatan Inc., Pleasanton, CA, USA). Specimens were observed with a TECNAI transmission electron microscope (Hillsboro, OR, USA) operating at 120–200 kV acceleration voltage equipped with a 4 k CCD camera FEI Eagle (Hillsboro, OR, USA).

2.3. Electrochemical Measurements

Electrochemical measurements were performed using a CH Instruments bipotentiostat model CHI750B (Austin, USA). The three-electrode system consisted of a glassy carbon electrode (GC macroelectrode A = 0.07 cm^2, GC microelectrode A = 98.4 μm^2), Ag, AgCl|3 M KCl_{aq} reference electrode, and Pt foil as a counter electrode. The GC electrodes were polished each time before measurement with alumina powder (1.0, 0.3, and 0.05 μm) on a polishing cloth. The electrodes were rinsed with ultrapure water, sonicated in an ultrasonic bath and left to dry in air. Then, the working space of the GCE was covered with a ca. 1 mm thin film of the cubic phase (which corresponded to ca. 5 mg of the phase). The measurements were performed at 25 °C and 37 °C using a BVT MT-1 minithermostat to maintain constant temperature. The electrochemical measurements were performed in a glass cell (BVT company, Brno, Czech Republic) with a volume of 10 mL. The solution in the cell was deoxygenated by passing high purity argon through the solution for 15 min before the measurement and the flow of pure gas over the solution was maintained during the experiment. We performed cyclic (CV), differential pulse (DPV) or square wave (SWV) voltammetry experiments on GC electrodes modified with a cubic phase. The CV experiments were carried out at various scan rates within the constant potential window of −0.2 V (starting point) to −1.2 V. For each type of cubic phase, the experiments were done in triplicate. The release profiles of MTX from the LCPs were obtained by recording the DPV peaks of MTX reduction. The pulse amplitude was 50 mV, pulse width 70 mV, and a pulse period 0.2 s. The saturation of the empty cubic phase with MTX was monitored also in the SWV mode with the following parameters: frequency 25 Hz, potential increment 2 mV, and amplitude 0.01 V.

2.4. Modeling of the Kinetics of Drug Release

There are several literature models that describe the drug release from different matrices [37]. In accordance with our previous work [27], to determine the kinetics of the drug elution from the LCPs, the Korsmeyer–Peppas model was chosen as it describes the controlled drug release from polymeric systems similar to cubic phases. The kinetics of the drug release, according to this model, are described by the following equation:

$$\frac{M_t}{M_\infty} = kt^n, \quad (1)$$

where M_t/M_∞ is the fraction of drug released at time t versus t_∞; k is the release rate constant; and the n value gives details on the release mechanism of the drug. To determine the value of n, the logarithm of less than the initial 60% of the release data values were fitted versus the log time. In the case of a cylindrical surface of a drug-containing carrier, when $n = 0.45$, the diffusion mechanism corresponds to Fick's law; for n between 0.45 and 0.89, the transport is non-Fickian; $n = 0.89$ corresponds to Case-II transport; and $n > 0.89$ for Super Case-II transport [37].

When $n \approx 0.5$, the Korsmeyer–Peppas approach is simplified to the Higuchi model, which describes the drug release from semi-solid matrices. The Higuchi model is described by the following expression:

$$M_t/M_\infty = k_H\sqrt{t}, \quad (2)$$

where k_H is the Higuchi constant.

The drug release profiles from the cubic mesophase films were evaluated from measurements performed using both spectrophotometry and voltammetry. The former reflects the increasing concentration of the drug in the solution while electrochemistry shows the removal from the films present at the electrode monitored by the underlying electrode. The spectrophotometry approach is more common, however, the electrochemical approach shows the release from the film and any solution phenomena such as the adsorption of the released drug on glassware will not affect the result. In the case of magnetocubosomes, only UV-Vis spectrophotometry was employed.

2.5. Magnetic Field Generator

The magnetic field generator was a custom-made inductor coil connected to an alternating electric power outlet (20 V, 50 Hz). The void of the inductor (stator) coil provided space for a quartz cuvette with a suspension of magnetocubosomes. Due to the design of the set-up, the magnetic field inside the inductor and quartz cuvette was not isotropic and its maximum value was measured as 10 mT. A constant temperature was maintained (25 °C) during the release experiments and the coil was cooled using a fan [32].

2.6. Spectroscopic Measurements

The UV-Vis measurements were also used to drug elution studies. The GCE was modified with a thin film of cubic phase containing 1.00% MTX and was placed in the cell filled with 10 mL of 0.1 M phosphate buffer at 25 °C. The cell was placed on a magnetic stirrer with gently stirring to keep the same concentration of drug in the entire volume of the cell. After a specified time, 2 mL of the solution was transferred to an acrylic cuvette and placed in the spectrophotometer. After the measurement the solution was returned to the cell. Experiments showing AMF-stimulated drug release from hybrid materials were performed using a UV-Vis Cary 60 spectrophotometer (Agilent Technologies, Warsaw Poland) with a 1 cm quartz cuvette in the wavelength range from 600 to 250 nm. A series of samples, each containing 0.5 mg/mL magnetocubosome dispersion in 0.1 M phosphate buffer at pH 7.4, was prepared. Each of the samples was subsequently placed in the cuvette inside the inductor coil for a specified time. Then, the cuvette was removed from the inductor coil and the absorbance was measured at 303 nm. For each experiment, a fresh sample was used.

The spectroscopic measurements were also used to calculate the entrapment efficiency (%) of MTX in magnetocubosomes. The freshly prepared magnetocubosomes were diluted ten times. Then, the diluted sample was placed in an Amicon Ultra-0.5 centrifugal filter unit with an Ultracel-3 membrane (NMWL 3 kDa, Sigma Aldrich, Munich, Germany) and centrifuged for 20 min at 6000 rpm to separate the unbound drug. The separated aqueous phase with free drug was ten times diluted in 0.1 M phosphate buffer and the absorbance was measured at 303 nm. The entrapment efficiency (EE) was determined using the following expression:

$$EE = \frac{C_T - C_F}{C_T} \cdot 100\%, \tag{3}$$

where C_T is the concentration of the drug added to the cubosome during preparation and C_F is the concentration of the free drug in the ultrafiltrate detected after centrifugation.

The standard calibration curve for MTX solution in 0.1 M phosphate buffer at pH 7.4 was developed from at least nine dilutions in the concentration range of 20.0–0.89 μg/mL. The absorption spectra of the MTX buffer solutions were measured in the wavelength range of 600–250 nm with the characteristic λ at 303 nm. All measurements were performed at 25 °C.

3. Results and Discussion

3.1. Structural Characterization of the MTX-Doped Cubic Phases

It is well known that a monoolein/water mixture in a ratio of 60/40% w/w forms a bicontinuous cubic phase characterized by $Pn\overline{3}m$ symmetry [38–40]. The presence of MTX up to 2% w/w did not compromise the internal structure of the monoolein bicontinuous cubic phase. The size of the drug was smaller than the length of the unit cell of the cubic phase. The SAXS data of the pure MO phase and those doped with 2% and 1% w/w of MTX at 25 °C and 37 °C are shown in Figure 1. The peak positions of a cubic lattice were observed at both temperatures. These peak positions were consistent with the cubic phase of the $Pn\overline{3}m$ space group (characteristic Bragg diffractions: $\sqrt{2}$:$\sqrt{3}$:$\sqrt{4}\sqrt{6}$:$\sqrt{8}$:$\sqrt{9}$ are shown near the corresponding peaks in Figure 1). The calculated values of the lattice parameter, lipid length, and diameter of the water channels are listed in Table 1. The values were calculated from the SAXS spectra (for more details see S2 and Kulkarni et al. [22]). Incorporation of MTX into the monoolein phase did not affect the lattice parameter or the size of the water channels (Table 1), and noticeable differences of ca. 0.1 nm were on the verge of error.

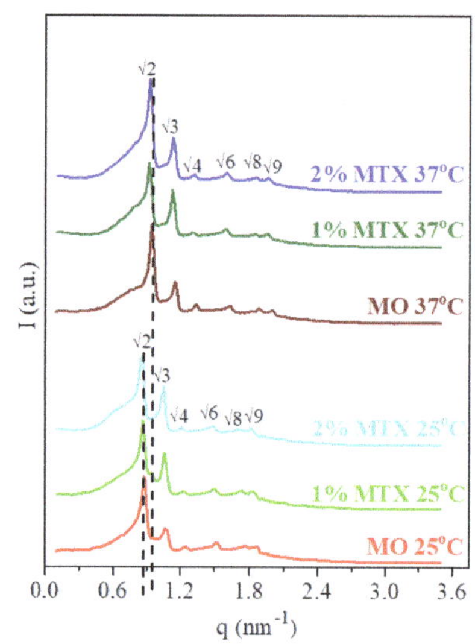

Figure 1. Comparison of SAXS spectra for the bulk cubic phases with 40% w/w of buffer and different amounts of MTX at 25 °C and 37 °C. Peaks are indexed for a double diamond cubic phase. The dashed lines show the position of the first peak for MO phase at 25 °C and 37 °C.

Table 1. Results of SAXS measurements for the monoolein cubic phase with and without the drug: phase symmetry, lattice parameter *a*, lipid length *l*, and water channel diameter d_w.

	T [°C]	Symmetry	a [nm]	l [nm]	d_w [nm]
MO/aq*	25	$Pn\overline{3}m$	10.2	1.7	4.5
60/40%	37	$Pn\overline{3}m$	9.5	1.6	4.2
MO/MTX/aq*	25	$Pn\overline{3}m$	10.3	1.8	4.5
59/1/40%	37	$Pn\overline{3}m$	9.8	1.7	4.3
MO/MTX/aq*	25	$Pn\overline{3}m$	10.3	1.8	4.6
58/2/40%	37	$Pn\overline{3}m$	9.7	1.6	4.3

*'aq' is used as an abbreviation for the phosphate buffer.

3.2. Electrochemical Measurements

3.2.1. Methotrexate Incorporated in the Monoolein Cubic Phase

MTX is an electroactive molecule that undergoes two redox processes: (1) The oxidation of MTX pyrazine moiety to 7-hydroxymethotrexate, which is an irreversible process occurring at positive potentials with the exchange of $2e^-/2H^+$; and (2) The reduction at negative potentials of the diamino-pteridinyl moiety to its 5,8-dihydro derivative via transfer of $2e^-/2H^+$, which is also reversible (Scheme S3) [41].

The electrochemical behavior of MTX was investigated using cyclic (CV), square wave (SWV), and differential pulse voltammetry (DPV) in 0.2 M phosphate buffer, pH 7.4. Previous reports have shown that the electrochemical response of MTX to CGE differs depending on the buffer pH [41,42]. Even though the obtained peaks were well resolved in acidic media, our main goal was to examine the drug's behavior at physiological pH. We began our investigations by examining the redox processes of MTX at an electrode covered with LCP saturated with the drug. This was achieved by dipping the electrode modified with monoolein LCP in a 10 mL, 2.33 mM solution of MTX in 0.2 M phosphate buffer at room temperature. The process of cubic phase saturation with MTX was monitored using the highly sensitive SWV method. After 15 min, the peak correlating to drug reduction was observed. Saturation was reached after 8 h (Figure 2). When a constant SWV peak current was attained, cyclic voltammograms were recorded at different scan rates. The dependencies of the reduction peak currents on the square root of the scan rate was linear (R^2 = 0.990), proving the diffusion control of the electrode processes (Figure 2 inset).

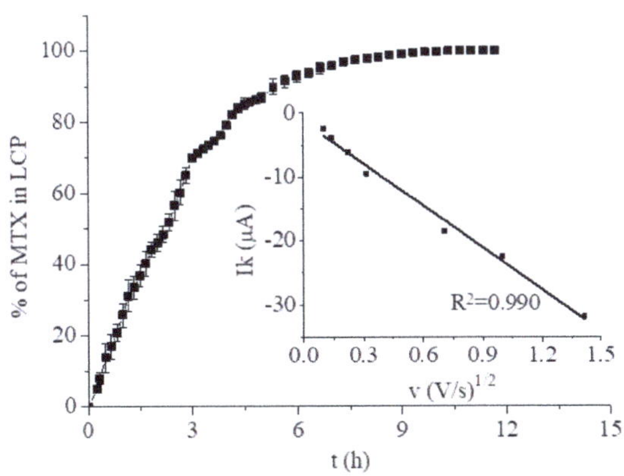

Figure 2. Saturation of LCP with the drug from 2.33 mM MTX solution in 0.2 M phosphate buffer monitored by SWV. Inset: Dependence of the CV reduction peak current at −0.790 V vs. the square root of the scan rate for the cubic phase layer phase with MTX after 12 h of saturation.

In the next step, MTX was incorporated into the monoolein cubic phase at several concentrations: 0.10, 0.25, 0.50, and 1.00% w/w. Drug-loaded cubic phases were used to modify the normal size GCE and GC microelectrodes. The working space of the electrodes was modified with a thin film of cubic phase (ca. 1 mm) containing MTX and transferred to the cell filled with 10 mL of deoxygenated phosphate buffer at room temperature. The response of the drug in negative potentials was monitored using DPV for the normal sized electrode and CV for the microelectrode. In the case of GCE, the transport to the electrode surface was determined by linear diffusion and recorded as a peak. The DPV reduction peak currents of the drug appeared ca. −0.790 V and increased with MTX concentration (Figure 3A). In the case of the microelectrode, the diffusion of the electroactive molecules was spherical and the CV response was in the form of a wave (Figure 3B). The wave height increased with the amount of MTX in the LCP. In both cases, the dependence of the MTX response on the drug concentration in the LCP was linear (Figure 3 inset).

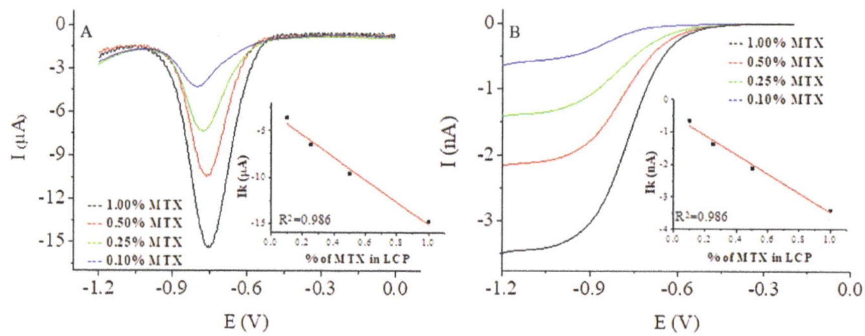

Figure 3. DPV recorded from a GCE (**A**) and CV from a GC microelectrode (**B**) modified with the cubic phase containing 1.00, 0.50, 0.25, and 0.10% w/w of MTX at pH 7.4. Inset: Dependence of the MTX response on the drug concentration in the LCP.

Drug elution studies of MTX from monoolein cubic phases were conducted at several concentrations: 0.25, 0.50, and 1.00% w/w. The spectroscopy and voltammetry measurements were performed under an argon atmosphere at 25 °C and 37 °C. The drug release profiles were investigated based on the changes in the DPV voltammograms with time counting from the moment of immersion of the electrodes in the pristine buffer are shown in Figure 4. Over the course of time, the peak current decreased, which directly shows the elution of the drug. The dependence of the peak current on measurement time is shown in Figure S4. The results were plotted as the normalized current (I/Io) vs. time and are presented in Figure 4. Figure 4A demonstrates the elution of the drug at room temperature. The profiles of the drug release were similarly independent of the initial concentration of the drug in the cubic phase. In all cases, the total elution was reached after ca. 350 min, and 50% w/w MTX (T_{50}) was released after 42 min for 0.25 and 0.50% w/w MTX and after 40 min for 1.00% w/w MTX. Figure 4B shows the elution of MTX at 37 °C, with the total elution of the drug being achieved after 250 min. T_{50} occurred after ca. 32 min for 0.25 and 0.50% MTX and 29 min for 1.00% MTX. These results can be attributed to the faster moving molecules at higher temperatures.

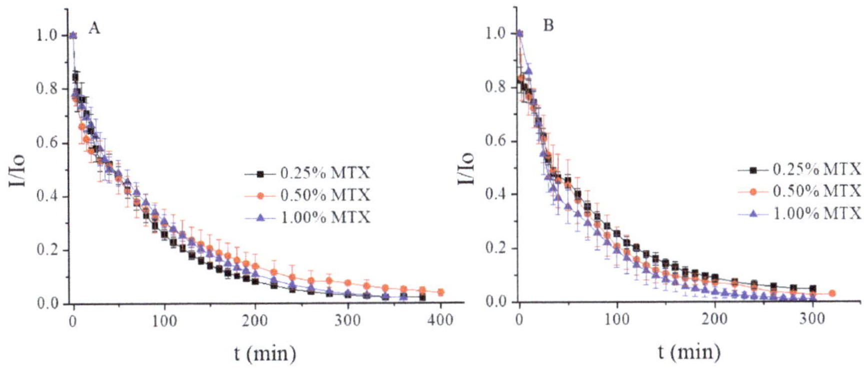

Figure 4. Release profile of MTX from LCPs at pH 7.4 at 25 °C (**A**) and 37 °C (**B**).

For comparison, we measured the elution of 1.00% w/w of MTX from LCP using UV-Vis at 25 °C, keeping the same conditions as in the electrochemical measurements. The drug release profile was investigated based on the increasing concentration of the drug in the solution surrounding the GCE modified with cubic phase. The comparison of release profile of 1% MTX measured with spectroscopy and voltammetry is presented in Figure 5. The profiles of the drug elution were similar for the first 60 min and after 200 min. The total elution was achieved at the same time, but the T_{50} measured with UV-Vis was achieved after ca. 30 min, 10 min earlier than using the electrochemical approach.

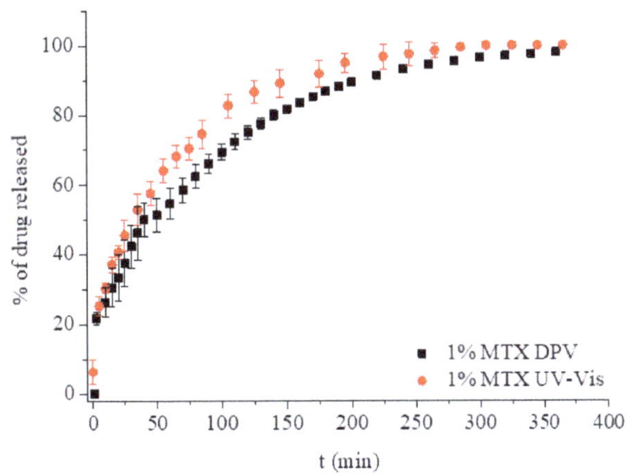

Figure 5. Release profiles of MTX from LCPs measured with UV-Vis spectroscopy and differential pulse voltammetry at pH 7.4 at 25 °C.

All of the drug release data were plotted as a log percentage of the drug release vs. log time [43,44]. The slope of the plot allowed for the evaluation of the exponent n in the Korsmeyer–Peppas model. The results obtained for MTX obtained using DPV and UV-Vis approaches at 25 °C gave an n value close to 0.45, which means that the diffusion mechanism corresponded to Fick's laws and the release mechanism proceeded according to the Higuchi model. With increasing temperature, the n value in all cases started to deviate from 0.45, indicating the influence of the LCP matrix and a transition to non-Fickian transport (see Table 2).

Table 2. Release characteristics of methotrexate from LCPs.

	Korsmeyer–Peppas			Higuchi	
% MTX	n	R^2	k [%/h^n]	k_H [%/h]	R^2
1.0 at 25 °C	0.44 ± 0.05	0.986 ± 0.004	56.08 ± 5.40	47.49 ± 1.49	0.977 ± 0.010
1.0 at 25 °C CUV-Vis	0.44 ± 0.01	0.998 ± 0.001	61.20±1.92	58.78 ± 1.32	0.997±0.001
1.0 at 37 °C	0.65 ± 0.19	0.921 ± 0.102	66.36 ± 6.32	a	a
0.5 at 25 °C	0.41 ± 0.03	0.967 ± 0.006	62.34 ± 2.22	40.51 ± 9.49	0.943 ± 0.038
0.5 at 37 °C	0.56 ± 0.11	0.950 ± 0.043	61.38 ± 8.76	66.22 ± 4.21	0.966 ± 0.020
0.5 and 2% MNPs at 25 °C	0.47 ± 0.06	0.949 ± 0.014	60.60 ± 2.73	62.68 ± 5.85	0.945 ± 0.021
0.25 at 25 °C	0.45 ± 0.08	0.975 ± 0.005	57.12 ± 3.80	53.97 ± 3.38	0.982 ± 0.008
0.25 at 37 °C	0.57 ± 0.09	0.943 ± 0.014	63.51 ± 4.18	72.21 ± 4.11	0.962 ± 0.009

a: not determined.

3.2.2. Behavior of MTX Incorporated into Hybrid LCP Systems

We have previously reported that the incorporation of up to 2% w/w of hydrophobic magnetic nanoparticles (MNPs) had no effect on the phase symmetry [27]. Due to their size (ca. 7 nm) and lipid coating, MNPs are located mainly in the lipidic domain of the LCP while the water channels remain free. Therefore, the magnetic nanoparticles do not block access of the drug to the electrode surface. A hybrid phase with MTX led to a similar outcome. MTX release profiles from the monoolein hybrid cubic phase containing 2% w/w of MNPs and 0.5% w/w of drug were monitored using the DPV method (Figure 6 and Figure S5). The measurements were carried out at 25 °C under an argon atmosphere. Incorporation of 2% w/w MNPs resulted in an increase of the initial peak current of MTX by 4 μA, without a change in the peak potential, −0.790 V (Figure S5). In the presence of MNPs, the mechanism of the drug release at room temperature was similar to the MO phase. The total elution of MTX was maintained at the same level as the pure monoolein phase. In this case, the value of T_{50} was 35 min and the n value of ca. 0.45 suggests that the drug was released according to Fick's law (see Table 2).

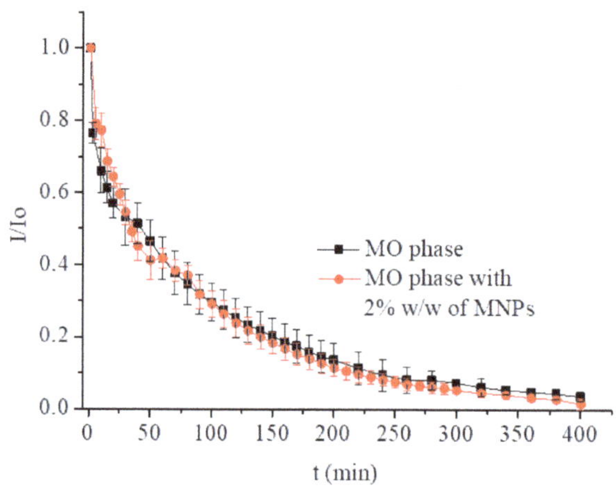

Figure 6. Release profiles of MTX from LCPs with and without magnetic nanoparticles at pH 7.4 at 25 °C.

3.2.3. MTX Incorporated into Magnetocubosomes

Nanoparticles of the hybrid cubic phase, magnetocubosomes, can be used for magnetic drug targeting, hyperthermia treatment, or magnetic resonance imaging [40] due to their small size and magnetic properties that allow them to move in a magnetic field. When loaded with the appropriate anti-cancer drug such as MTX, they can serve as a platform for drug targeting delivery systems. Cubosomes and magnetocubosomes loaded with MTX were prepared, and their characteristics evaluated. The cubic nature of the LCP nanoparticles was characterized by SAXS. The SAXS pattern for cubosomes with MTX showed the relative positions of the diffraction signals—$\sqrt{2}$:$\sqrt{4}$:$\sqrt{6}$—corresponding to the $Im\overline{3}m$ space group (Figure 7, black line). In the case of the magnetocubosomes, identical signals were observed (Figure 7, red line). The crystallographic lattice parameter was 13.9 nm and 14.3 nm for the cubosomes and magnetocubosomes, respectively. The diameter of the magnetocubosomes doped with MTX was evaluated using DLS measurements, which was ca. 130 nm, and the zeta potential was ca. −30 mV (Figure S6). The entrapment efficiency of MTX in the magnetocubosomes was evaluated using Equation (3) and a standard calibration curve for MTX solution in 0.1 M phosphate buffer at pH 7.4 (Figure S7). The load efficiency was 63.1%.

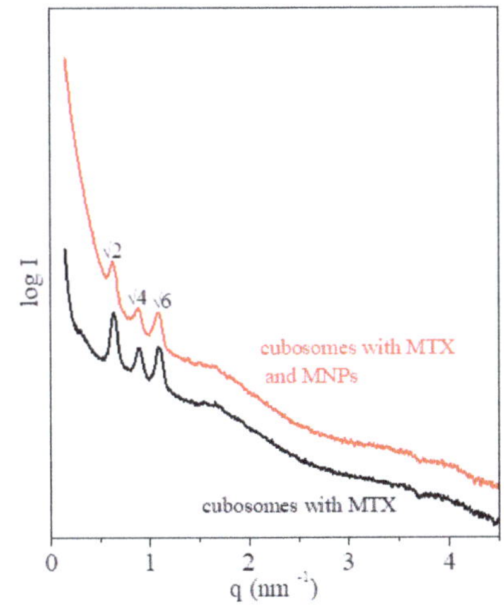

Figure 7. SAXS spectra of the cubosomes and magnetocubosomes with methotrexate at room temperature (25 °C).

Cryo-TEM allows for the imaging of the cubosomes without any changes of their structure. The Cryo-TEM images for cubosomes/magnetocubosomes with MTX showed a well-ordered structure inside the particles (Figure S8). This highly ordered internal structure of the magnetocubosome provides an extensive hydrophobic–hydrophilic interfacial area, e.g., for the solubilization of different drugs and offers a convenient way of delivering the drug into the body. Additionally, the MNPs were visible inside the magnetocubosomes; hence the magnetocubosomes were expected to be sensitive towards magnetic fields. In the Supplementary Video S9 a link to a film showing the movement of magnetocubosomes in the field of the magnet can be found.)

3.3. Low-Frequency Alternating Magnetic Field (AMF)-Stimulated Drug Release

The effect that magnetic nanoparticles have on drug release as well as other molecules from various lipidic structures under the influence of the external magnetic field has been well documented. Vallooran et al. [30] showed that under a constant magnetic stimulus, the hexagonal phase with MNPs aligns the orientation of the hexagonal domains in the direction of the external (constant) magnetic field, thus facilitating the transport of a hydrophilic drug across the liquid crystalline phase in this direction. Two cases were considered: when the magnetically induced orientation of the hexagonal domains in the membrane was in the direction of diffusion, it facilitated transport of a hydrophilic drug, and when the forced position of the hexagonal domains was perpendicular to the diffusion direction, it resulted in slower drug diffusion. Application of LF-AMF increased the release of the hydrophilic molecules from the water channels of magnetocubosomes [33]. Mendozza et al. showed that the monoolein cubic phase with MNPs subjected to AMF for 10 min undergoes transition from $Pn\overline{3}m$ to H_{II} [45].

Other reports [31,33,35,46,47] relied on the so-called magnetothermal effect observed in an alternating magnetic field. This effect reveals itself as temperature rises, transferring heat to the immediate vicinity when the magnetic nanoparticles are exposed to the alternating magnetic field of sufficiently high frequency (typically from several kHz to hundreds of kHz) due to the Néel and Brownian relaxation [48]. This heat transfer affects the physicochemical properties of the lipidic structures, triggering drug release from lipid mesophases [30,33]. Since the heating efficiency, a prerequisite for the magnetothermal effect, varies as $\omega^2\tau/(1 + \omega^2\tau^2)$, where ω is the magnetic field frequency and τ is the effective relaxation time, under the LF-AMF conditions used in this work (50 Hz AMF, 10 mT, 20 V), the relaxation time was only ca. 10^{-6} s [49], therefore the magnetothermal effect as the main stimulus for drug release has to be excluded. Another mechanism of triggered drug release includes changes of the nanomechanical properties of the hybrid MNP–lipidic material. The incorporation of oleic acid-coated nanoparticles into the solid-supported or the free-standing BLMs resulted in an increase of the Young's modulus of elasticity measured by the nanoidentation and electrostriction methods [32]. This modulus is a measure of stiffness of the lipid bilayer and its increased value suggests that the membrane is prone to mechanical rupture in an alternating magnetic field by the vibrating superparamagnetic nanoparticles localized inside the bilayer walls. Even though the hydrophobic nanoparticles used (ca. 7 nm in diameter) seemed to be too large to be hosted fully in the lipid bilayer, which is 3–4 nm thick, the same explanation can be proposed here [48,49]. Earlier, the rupture of magnetoliposomes was confirmed by TEM and the doxorubicin outflow from magnetoliposomes was also shown under LF-AMF of 50 Hz frequency and 10 mT magnetic field intensity [32]. Therefore, we used this approach to induce the methotrexate release from magnetocubosomes, assuming that the mechanical vibration of magnetic nanoparticles rather than the localized heating of the MNPs' vicinity would lead to similar local disruption of magnetocubosomes and MTX release under LF-AMF. A freshly prepared sample of 0.5 mg/mL magnetocubosome dispersion was placed inside the inductor coil. After a specified time, the absorbance was measured at 303 nm (Figure 8). The application of AMF led to an increase in the rate of drug release from the cubosomes. The relatively large data scatter, regardless of the number of repetitions, was ascribed to the ex situ approach, where the sample was transferred from the inductor coil to the spectrophotometer chamber. Nevertheless, two important facts can be extrapolated from our results. First, the release takes place very fast and reaches saturation within the first 5 min of the application of AMF. This release appears to be much faster than in the case of magnetoliposomes [32]. After this time, the amount of drug released from the magnetocubosomes reached 2.6 μg/mg (30% of the drug released when compared to the initial drug load). Second, over the same time period, the release from magnetocubosomes without AMF stimulus was only ca. 0.16 μg/mg, hence 2% of the drug released when compared to the initial drug load. The dashed curve shown in Figure 8, even though obtained by fitting the results to the Korsmeyer-Peppas model for magnetocubosomes stimulated with AMF, is the possible outcome. Nevertheless, it is clear that the application of LF-AMF triggers a larger and faster release of the drug from the magnetocubosomes. Furthermore, since the frequency of the AMF stimulus and the relaxation time of the magnetic nanoparticles used were insufficient to trigger the magnetothermal effect (vide supra and reference [32]), we presumed that the alternating magnetic field at a low frequency of 50 Hz caused mechanical vibration of the magnetic nanoparticles that can rupture to some extent the internal structure of magnetocubosomes, facilitating the drug's delivery from the aqueous channels of the cubosomes. Larger disturbances may lead to changes in the internal structure and possible phase transition [45].

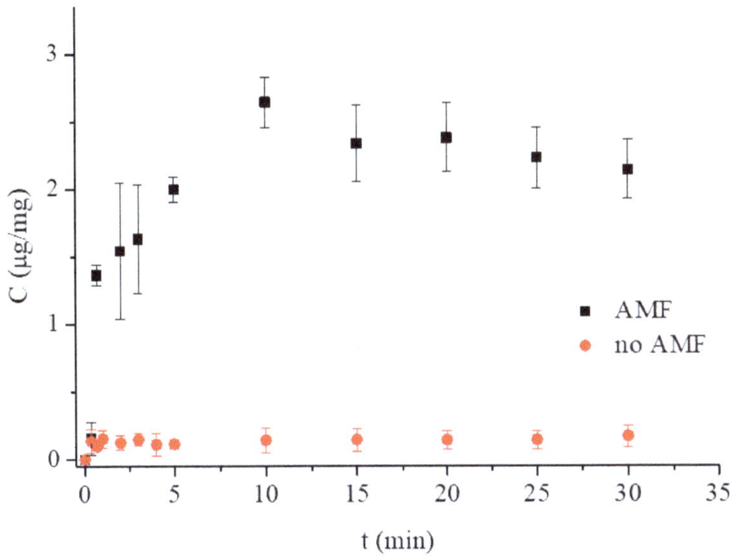

Figure 8. MTX released from magnetocubosomes vs. time, spontaneously and caused by LF-AMF at pH 7.4 and 25 °C.

4. Conclusions

Methotrexate is a widely used drug; however, due to its toxic side effects, it should be delivered to organisms encapsulated using an appropriate drug delivery vehicle to prevent healthy cells from being exposed to its toxic influence. We have shown that the lipidic liquid crystalline cubic phase can be a promising matrix for such a task. The release profiles of the drug from the films of the cubic phases placed at electrodes are described by the Higuchi model. The removal of the drug from the cubic phase was facilitated by an increase in temperature. The cubic phase dispersed into cubosomes should be even more attractive as MTX nanocarriers since they retain the internal structure of the parent bulk phase while the viscosity of the system is reduced, which makes it easier to apply.

Hydrophobic magnetic nanoparticles could be easily incorporated into the cubic phase nanoparticles. Since cubosomes are made of lipids organized in very expanded, curved bilayers they should be prone to the mechanical vibration of numerous magnetic nanoparticles present in these bilayers. Joniec et al. observed an increase in the Young's modulus due to the presence of hydrophobic magnetic nanoparticles in magnetoliposomes loaded with doxorubicin [32]. The rupture of magnetoliposomes placed in the low-frequency alternating magnetic field due to mechanical vibration was observed by TEM and the release of doxorubicin from magnetoliposomes in the LF-AMF of 50 Hz frequency and 10 mT magnetic field intensity was monitored with fluorescence and voltammetry. Methotrexate release from the magnetocubosomes placed in the LF-AMF of the same frequency and field intensity lead also to local disruption of magnetocubosomes and release of MTX, as confirmed with UV-Vis spectroscopy. After the same length of time, MTX was barely released from magnetocubosomes in the absence of LF-AMF, which confirmed the role of magnetic field in triggering drug release from the cubosome.

These results show that magnetocubosomes are interesting hybrid nanostructures that may accommodate a suitable amount of methotrexate, due to the structure—larger than other carriers. The rate and efficiency of methotrexate release at the appropriate sites may be controlled, e.g., by applying an alternating magnetic field. Magnetic field can be used to relocate the dispersion of magnetocubosomes without any leakage of the magnetic nanoparticles from the cubosome lipid phase to the solution. The new hybrid dispersed nanomaterial is certainly promising as a carrier and requires in vitro and in vivo studies. Leesajakul et al. [50] studied the stability of cubosomes with Pluronic 127 as the stabilizer in plasma and reported that HDL, LDL, and albumin interacted with cubosome particles. The destabilization of cubosomes in plasma resulted in smaller particle remnants These authors suggested that long circulation time of an incorporated substance in cubosomes was due to the sustained behavior of cubosomes' remnant particles. In a recent paper, Tran et al. [51] confirmed the accumulation of both cubosomes and hexosomes in the livers and spleens of mice up to 20 h post-injection. Their biodistribution and histology analysis results indicated that cubosomes and hexosomes should be considered as candidates for macrophage-associated theranostic systems. The interactions of cubosomes with proteins were also studied by, e.g., Falchi et al. [52] and Biffi et al. [53]. These authors show that the stability of cubosomes in the body fluids depends on the lipid and especially polymer used in the in vivo studies. Pluronic F108 and PEG were found to be the best stabilizers of cubosomes and hexosomes in the organism. We presented in vitro cytotoxicity studies for monoolein cubosomes and hexosomes containing other drugs [54–56]), confirming the stability of these nanoparticles in the cell environment. Further in-depth studies are needed to prove their utility for medical applications.

Supplementary Materials: The following are available online at http://www.mdpi.com/2079-4991/9/4/636/s1, Figure S1: Scheme of chemical structure of methotrexate (MTX) and folic acid (FA). S2: Equations used for the determination of Symmetry, lattice parameter, lipid length and diameter of water channels of the cubic phases. Figure S3: Scheme of the reduction process of methotrexate. Figure S4: Release profiles of MTX from a cubic phase in pH 7.4 at 25 and 37 °C. Figure S5: DPV on GC electrode modified with phases without and with magnetic nanoparticles and the release profiles of MTX from LCPs at pH 7.4 at 25 °C. Figure S6: The size and zeta potential of magnetocubosomes containing MTX determined with DLS at 25 °C. Figure S7: Standard calibration curve for methotrexate based on measurement at 303 nm in 0.1 M phosphate buffer, pH 7.4. Figure S8: Electron cryo-microscopy images of cubosomes and magnetocubosome. Video S9: Movement of magnetocubosomes in magnetic field.

Author Contributions: Equal contribution of all authors to the work reported is declared

Funding: This research was funded by the National Science Centre, conferred on the basis of decision number DEC-2016/23/B/ST4/03295.

Acknowledgments: This project was financed by the National Science Centre, conferred on the basis of decision number DEC-2016/23/B/ST4/03295. MM thanks Sinergia SNSF project no. CRSII2_154451 for support. CIISB research infrastructure project LM2015043 funded by MEYS CR is gratefully acknowledged for their financial support of the measurements at the CF Cryo-electron Microscopy and Tomography CEITEC MU. The authors would like to thank Dr. Jiri Novacek from CIISB for the Cryo-TEM measurements and Aleksandra Rękorajska, University of Warsaw, for the synthesis of the magnetic nanoparticles. The authors would like to thank Dr. hab. Damian Pociecha, University of Warsaw, for help in SAXS measurements of the cubosomes.

Conflicts of Interest: The authors declare no conflict of interest.

Abbreviations

MTX	methotrexate
SAXS	small angle X-ray scattering
LCP	liquid crystalline phase
DDS	drug delivery system
AMF	alternating magnetic field
MO	monoolein
MNPs	magnetic nanoparticles
DLS	dynamic light scattering
Cryo-TEM	cryogenic transmission electron microscopy
DPV	differential pulse voltammetry
GC	glassy carbon
GCE	glassy carbon electrode
SWV	square-wave voltammetry
CV	cyclic voltammetry
EE	entrapment efficiency

References

1. Genestier, L.; Paillot, R.; Quemeneur, L.; Izeradjene, K.; Revillard, J.-P. Mechanisms of action of methotrexate. *Immunopharmacology* **2000**, *47*, 247–257. [CrossRef]
2. Bleyer, W.A. The clinical pharmacology of Methotrexate. *Cancer* **1978**, *41*, 36–51. [CrossRef]
3. Yoon, S.-A.; Choi, J.R.; Kim, J.-O.; Shin, J.-Y.; Zhang, X.H.; Kang, J.-H. Influence of Reduced Folate Carrier and Dihydrofolate Reductase Genes on Methotrexate-Induced Cytotoxicity. *Cancer Res. Treat.* **2010**, *42*, 163–171. [CrossRef]
4. Kanarek, N.; Keys, H.R.; Cantor, J.R.; Lewis, C.A.; Chan, S.H.; Kunchok, T.; Abu-Remaileh, M.; Freinkman, E.; Schweitzer, L.D.; Sabatini, D.M. Histidine catabolism is a major determinant of methotrexate sensitivity. *Nature* **2018**, *559*, 632–655. [CrossRef] [PubMed]
5. Rozensza, L.A.; Radnay, J. The effect of methotrexate on transformation and mitosis of normal human-blood lymphocytes in-vitro. *Blood* **1974**, *43*, 401–409.
6. Bookbinder, S.A.; Espinoza, L.R.; Fenske, N.A.; Germain, B.F.; Vasey, F.B. Methotrexate therapy in the rheumatic disease. *Clin. Exp. Rheumatol.* **1984**, *2*, 185–193.
7. Zeb, A.; Qureshi, O.S.; Kim, H.-S.; Cha, J.-H.; Kim, H.-S.; Kim, J.-K. Improved skin permeation of methotrexate via nanosized ultradeformable liposomes. *Int. J. Nanomed.* **2016**, *11*, 3813–3824. [CrossRef]
8. Srinivas, P.S.; Babu, D.R.S. Formulation and evaluation of parenteral methotrexate nanoliposomes. *Int. J. Pharm. Pharm. Sci.* **2014**, *6*, 295–300.
9. Rosenholm, J.M.; Peuhu, E.; Bate-Eya, L.T.; Eriksson, J.E.; Sahlgren, C.; Linden, M. Cancer-cell-specific induction of apoptosis using mesoporous silica nanoparticles as drug-delivery vectors. *Small* **2010**, *6*, 1234–1241. [CrossRef]
10. de Oliveira Freitas, L.B.; Gonzalez Bravo, I.J.; de Almeida Macedo, W.A.; de Sousa, E.M.B. Mesoporous silica materials functionalized with folic acid: Preparation, characterization and release profile study with methotrexate. *J. Sol-Gel Sci. Technol.* **2016**, *77*, 186–204. [CrossRef]
11. Kakkar, D.; Dumoga, S.; Kumar, R.; Chuttani, K.; Mishra, A.K. PEGylated solid lipid nanoparticles: Design, methotrexate loading and biological evaluation in animal models. *Med. Chem. Commun.* **2015**, *6*, 1452–1463. [CrossRef]
12. Ferreira, M.; Silva, E.; Barreiros, L.; Segundo, M.A.; Costa Lima, S.A.; Reis, S. Methotrexate loaded lipid nanoparticles for topical management of skin-related diseases: Design, characterization and skin permeation potential. *Int. J. Pharm.* **2016**, *512*, 14–21. [CrossRef]
13. Hashada, R.A.; Ishaka, R.A.H.; Geneidia, A.S.; Mansoura, S. Methotrexate loading in chitosan nanoparticles at a novel pH: Response surface modeling, optimization and characterization. *Int. J. Biol. Macromol.* **2016**, *91*, 630–639. [CrossRef] [PubMed]
14. Pretti, T.S.; Souza, M.A.; Santos, H.T.; Nascimento, A.C.G.G.; Santos, F.J.; Fraga, A.F.; Jafelicci, M.J.; Marques, R.F.C. Drug delivery nanotechnology applied to Methotrexate controlled release in human osteosarcona: In vitro test. Available online: http://www.metallum.com.br/7colaob/resumos/trabalhos_completos/08-020.docx (accessed on 15 April 2019).
15. Guo, F.; Fan, Z.; Yang, J.; Li, Y.; Wang, Y.; Zhao, H.; Xie, L.; Hou, Z. A comparative evaluation of hydroxycamptothecin drug nanorods with and without methotrexate prodrug functionalization for drug delivery. *Nanoscale Res. Lett.* **2016**, *11*, 384–396. [CrossRef] [PubMed]
16. Pereira, A.D.F.; Pereira, L.G.R.; Barbosa, L.A.D.O.; Fialho, S.L.; Pereira, B.G.; Patricio, P.S.D.O.; Pinto, F.C.H.; Da Silva, G.R. Efficacy of methotrexate-loaded poly(e-caprolactone) implants in Ehrlich solid tumor-bearing mice. *Drug Deliv.* **2013**, *2*, 168–179. [CrossRef] [PubMed]
17. Ghorbani, M.; Hamishehkar, H.; Arsalani, N.; Entezami, A.A. A novel dual-responsive core-crosslinked magnetic-gold nanogel for triggered drug release. *Mater. Sci. Eng. C* **2016**, *68*, 436–444. [CrossRef]
18. Lina, L.; Xua, W.; Lianga, H.; Hea, L.; Liua, S.; Lia, Y.; Lia, B.; Chena, Y. Construction of pH-sensitive lysozyme/pectin nanogel for tumor methotrexate delivery. *Colloid. Surface. B* **2015**, *126*, 459–466. [CrossRef] [PubMed]
19. Guo, C.; Wang, J.; Cao, F.; Lee, R.J.; Zhai, Z. Lyotropic liquid crystal systems in drug deliver. *Drug Discov. Today* **2010**, 1032–1040. [CrossRef]

20. Angelova, A.; Garamus, V.M.; Angelov, B. Advances in structural design of lipid-based nanoparticle carriers for delivery of macromolecular drugs, phytochemical and anti-tumor agents. *Adv. Colloid Interf. Sci.* **2017**, *249*, 331–345. [CrossRef]
21. Barriga, H.M.G.; Holme, M.N.; Stevens, M.M. Cubosomes: The next generation of smart lipid nanoparticles? *Angew. Chem. Int. Ed.* **2018**, *57*, 2–23. [CrossRef]
22. Kulkarni, C.V.; Wachter, W.; Iglesias-Salto, G.; Engelskirchenb, S.; Ahualliac, S. Monoolein: A magic lipid? *Phys. Chem. Chem. Phys.* **2011**, *13*, 3004–3021. [CrossRef]
23. Nazaruk, E.; Szlęzak, M.; Górecka, E.; Bilewicz, R.; Osornio, Y.M.; Uebelhart, P.; Landau, E.M. Design and Assembly of pH-Sensitive Lipidic Cubic Phase Matrices for Drug Release. *Langmuir* **2014**, *30*, 1383–1390. [CrossRef] [PubMed]
24. Negrini, R.; Sanchez-Ferrer, A.; Mezzenga, R. Influence of Electrostatic Interactions on the Release of Charged Molecules from Lipid Cubic Phases. *Langmuir* **2014**, *30*, 4280–4288. [CrossRef]
25. Negrini, R.; Fong, W.-K.; Boyd, B.J.; Mezzenga, R. pH-responsive lyotropic liquid crystals and their potential therapeutic role in cancer treatment. *Chem. Commun.* **2015**, *51*, 6671–6674. [CrossRef] [PubMed]
26. Negrini, R.; Mezzenga, R. pH-Responsive Lyotropic Liquid Crystals for Controlled Drug Delivery. *Langmuir* **2011**, *27*, 5296–5303. [CrossRef]
27. Szlezak, M.; Nieciecka, D.; Joniec, A.; Pękała, M.; Gorecka, E.; Emo, M.; Stébé, M.J.; Krysiński, P.; Bilewicz, R. Monoolein cubic phase gels and cubosomes doped with magnetic nanoparticles - hybrid materials for controlled drug release. *ACS Appl. Mater. Inter.* **2017**, *9*, 2796–2805. [CrossRef]
28. Fong, W.-K.; Negrini, R.; Vallooran, J.J.; Mezzenga, R.; Boyd, B.J. Responsive self-assembled nanostructured lipid systems for drug delivery and diagnostics. *J. Colloid Interf. Sci.* **2016**, *484*, 320–339. [CrossRef]
29. Bonini, M.; Berti, D.; Baglioni, P. Nanostructures for magnetically triggered release of drugs and biomolecules. *Curr. Opin. Colloid Interface Sci.* **2013**, *18*, 459–467. [CrossRef]
30. Vallooran, J.J.; Negrini, R.; Mezzenga, R. Controlling anisotropic drug diffusion in lipid-Fe_3O_4 nanoparticle hybrid mesophases by magnetic alignment. *Langmuir* **2013**, *29*, 999–1004. [CrossRef]
31. Monnier, C.A.; Burnand, D.; Rothen-Rutishauser, B.; Lattuada, M.; Petri-Fink, A. Magnetoliposomes: Opportunities and challenges. *Eur. J. Nanomed.* **2014**, *6*, 201–215. [CrossRef]
32. Joniec, A.; Sek, S.; Krysinski, P. Magnetoliposomes as Potential Carriers of Doxorubicin to Tumours. *Chem.-Eur. J.* **2016**, *22*, 17715–17724. [CrossRef]
33. Montis, C.; Castroflorio, B.; Mendozza, M.; Salvatore, A.; Berti, D.; Baglioni, P. Magnetocubosomes for the delivery and controlled release of therapeutics. *J. Colloid Interf. Sci.* **2015**, *449*, 317–326. [CrossRef] [PubMed]
34. Hong, S.K.; Kim, J.C. Cubic phase magnetic nanoparticles. *Mol. Cryst. Liq. Cryst.* **2015**, *607*, 123–134. [CrossRef]
35. Wang, M.H.; Kim, J.-C. Magnetic field-responsive cubosomes containing magnetite and poly(N-isopropylacrylamide). *J. Controlled Release* **2013**, *172*, e139. [CrossRef]
36. Ivkov, R.; DeNardo, S.J.; Daum, W.; Foreman, A.R.; Goldstein, R.; Nemkov, V.S.; DeNardo, G.L. Application of High Amplitude Alternating Magnetic Fields for Heat Induction of Nanoparticles Localized in Cancer. *Clin. Cancer Res.* **2005**, *11*, 7093–7103. [CrossRef] [PubMed]
37. Costa, P.; Sousa Lobo, J.M. Modeling and comparison of dissolution profiles. *Eur. J. Pharm. Sci.* **2001**, *13*, 123–133. [CrossRef]
38. Hyde, S.T.; Andersson, S.; Ericsson, B.; Larsson, K. A Cubic Structure Consisting of a Lipid Bilayer Forming an Infinite Periodic Minimum Surface of the Gyroid Type in the Glycerolmonooleat-Water System. *Z. Kristallogr.* **1984**, *168*, 213–219. [CrossRef]
39. Briggs, J.; Chung, H.; Caffrey, M. The Temperature-Composition Phase Diagram and Mesophase Structure Characterization of the Monoolein/Water System. *J. Phys.* **1996**, *6*, 723–751. [CrossRef]
40. Qiu, H.; Caffrey, M. The Phase Diagram of the Monoolein/Water System: Metastability and Equilibrium Aspects. *Biomaterials* **2000**, *21*, 223–234. [CrossRef]
41. Pontinha, A.D.R.; Jorge, S.M.A.; Diculescu, V.C.; Vivan, M.; Oliveira-Brett, A.M. Antineoplasic Drug Methotrexate Redox Mechanism Using a Glassy Carbon Electrode. *Electroanalysis* **2012**, *24*, 917–923. [CrossRef]
42. Gao, L.; Wu, Y.; Liu, J.; Ye, B. Anodic voltammetric behaviors of methotrexate at a glassy carbon electrode and its determination in spiked human urine. *J. Electroanal. Chem.* **2007**, *610*, 131–136. [CrossRef]

43. Acharya, D.P.; Moffat, B.A.; Polyzos, A.; Waddington, L.; Coia, G.; Wright, D.K.; Wang, H.X.; Egan, G.F.; Muir, B.W.; Hartley, P.G. Cubic mesophase nanoparticles doped with superparamagnetic iron oxide nanoparticles: A new class of MRI contrast agent. *RSC Adv.* **2012**, *2*, 6655–6662. [CrossRef]
44. Nazaruk, E.; Miszta, P.; Filipek, S.; Landau, E.M.; Bilewicz, R. Lyotropic cubic phase gels for drug delivery: Tuning diffusion and sustained release from the mesophase. *Langmuir* **2015**, *31*, 12753–12761. [CrossRef] [PubMed]
45. Mendozza, M.; Montis, C.; Caselli, L.; Wolf, M.; Baglionia, P.; Berti, D. On the thermotropic and magnetotropic phase behavior of lipid liquid crystals containing magnetic nanoparticles. *Nanoscale* **2018**, *10*, 3480–3488. [CrossRef] [PubMed]
46. Reimhult, E. Nanoparticle-triggered release from lipid membrane vesicles. *New Biotechnol.* **2015**, *32*, 665–672. [CrossRef] [PubMed]
47. Bixner, O.; Reimhult, E. Controlled megnetosomes: Embedding of magnetic nanoparticles into membranes of monodisperse lipid vesicles. *J. Colloid Interf. Sci.* **2016**, *466*, 62–71. [CrossRef]
48. Shaghasemi, B.S.; Virk, M.M.; Reimhult, E. Optimization of Magneto-thermally Controlled Release Kinetics by Tuning of Magnetoliposome Composition and Structure. *Sci. Rep.* **2017**, *7*, 7474. [CrossRef]
49. Fortin, J.-P.; Gazeau, F.; Wilhelm, C. Intracellular heating of living cells through Néel relaxation of magnetic nanoparticles. *Eur. Biophys. J.* **2008**, *37*, 223–228. [CrossRef]
50. Leesajakul, W.; Nakano, M.; Taniguchi, A.; Handa, A. Interaction of cubosomes with plasma components resulting in the destabilization of cubosomes in plasma. *Colloid. Surfaces B* **2004**, *34*, 253–258. [CrossRef]
51. Tran, N.; Bye, N.; Moffat, B.A.; Wright, D.K.; Cuddihy, A.; Hinton, T.M.; Hawley, A.M.; Reynolds, N.P.; Waddington, L.J.; Mulet, X.; et al. Dual-modality NIRF-MRI cubosomes and hexosomes: High throughput formulation and in vivo biodistribution. *Mater. Sci. Eng. C* **2017**, *71*, 584–593. [CrossRef]
52. Falchi, A.M.; Rosa, A.; Atzeri, A.; Incani, A.; Lampis, S.; Meli, V.; Caltagirone, C.; Murgia, S. Effects of monoolein based cubosome formulations on lipid droplets and mitochondria of HeLa cells. *Toxicol. Res.* **2015**, *4*, 1025–1036. [CrossRef]
53. Biffi, S.; Andolfi, L.; Caltagirone, C.; Garrovo, C.; Falchi, A.M.; Lippolis, V.; Lorenzon, A.; Maco, P.; Meli, V.; Monduzzi, M.; et al. Cubosomes for in vivo fluorescence lifetime imaging. *Nanotechnology* **2017**, *28*, 055102. [CrossRef] [PubMed]
54. Nazaruk, E.; Majkowska, A.; Bilewicz, R. Lipidic Cubic-Phase Nanoparticles—Cubosomes for Efficient Drug Delivery to Cancer Cells. *ChemPlusChem* **2017**, *82*, 570–575. [CrossRef]
55. Nazaruk, E.; Majkowska, A.; Godlewska, M.; Salamonczyk, M.; Gawel, D. Electrochemical and biological characterization of lyotropic liquid crystalline phases—Retardation of drug release from hexagonal mesophases. *J. Electroanal. Chem.* **2018**, *813*, 208–215. [CrossRef]
56. Godlewska, M.; Majkowska-Pilip, A.; Stachurska, A.; Gawel, D.; Nazaruk, E. Voltammetric and biological studies of folate-targeted non-lamellar lipid mesophases. *Electrochim. Acta* **2019**, *299*, 1–11. [CrossRef]

nanomaterials

MDPI

Article

Shape Memory Polyurethane Biocomposites Based on Toughened Polycaprolactone Promoted by Nano-Chitosan

Arvind Gupta and Beom Soo Kim *

Department of Chemical Engineering, Chungbuk National University, Cheongju, Chungbuk 28644, Korea; myarvind2003@gmail.com

* Correspondence: bskim@chungbuk.ac.kr; Tel.: +82-43-261-2372

Received: 17 January 2019; Accepted: 4 February 2019; Published: 7 February 2019

Abstract: The distinctive ability to remember their original form after partial or complete deformation makes shape memory polymers remarkable materials for several engineering and biomedical applications. In the present work, the development of a polycaprolactone based toughened shape memory polyurethane biocomposite promoted by in situ incorporation of chitosan flakes has been demonstrated. The chitosan flakes were homogeneously present in the polymer matrix in the form of nanoflakes, as confirmed by the electron microscopic analysis and probably developed a crosslinked node that promoted toughness (*a* > 500% elongation at break) and led to *a* ~130% increment in ultimate tensile strength, as analyzed using a universal testing machine. During a tensile pull, X-ray analysis revealed the development of crystallites, which resulted from a stress induced crystallization process that may retain the shape and melting of the crystallites stimulating shape recovery (with *a* ~100% shape recovery ratio), even after permanent deformation. The biodegradable polyurethane biocomposite also demonstrates relatively high thermal stability (T_{max} at ~360 °C). The prepared material possesses a unique shape memory behavior, even after permanent deformation up to *a* > 500% strain, which may have great potential in several biomedical applications.

Keywords: chitosan; polycaprolactone; shape memory; stretchability; polyurethane; biocomposite

1. Introduction

The distinctive ability to remember their original form after partial or complete deformation makes the shape memory polymers (SMP) remarkable materials for several engineering and biomedical applications. Along with the widely known shape memory alloy, Nitinol (Nickel-Titanium Naval Ordnance Laboratory) [1], various polymers such as polyacrylates [2], polyimides [3], styrene-butadiene copolymers [4,5], polyurethane (PU) [6], polystyrene [7], etc., are known to have a shape memory ability on application of an external stimulus, such as pH, electrical, mechanical, hydration, magnetic, heat, light, etc. [8,9]. Among these polymers, polyurethane is found to be a versatile material due to its extraordinary characteristics, such as its relatively high toughness, light weight, tolerance to strain, high strength and elongation at break, abrasion resistance, low cost, and easy processability; polyurethane has applications in various domains, ranging from electronic packaging, food packaging, biomedical, aerospace, automobiles, textiles, energy, etc., to commodity applications [10,11].

Polyurethanes are a class of polymers with inherent two-phase separated domains due to a microphase inhomogeneity containing a urethane bond between an alternating polymer block of hard and soft segments, wherein diisocyanate is a hard segment connected to the flexible soft segments made of polyether or polyester diol chains. The presence of a hard segment is responsible for remembering the original shape, whereas the soft segment stores the energy for dissipation and provides the desired force, allowing polyurethane to return to its original form. Herein, the soft segments, i.e., polyol, can broadly be classified as carbonate based (i.e., polyhexamethylene carbonate diol) [12], ether based (i.e., polytetrahydrofuran diol) [13], and ester based (i.e., polycaprolactone diol) [14]. These conventional polyols are generally obtained from petroleum-based resources, and only some of them are biodegradable [15].

It is now recognized by the scientific community that the biopolymers, biobased and biodegradable polymers, have the potential to replace existing petroleum based non-biodegradable polymers and may thus be considered promising future materials [16]. These polymers include polylactic acid (PLA), polycaprolactone (PCL), polyglycolic acid (PGA), polyhydroxybutyrate (PHB), etc., which have been found to be useful in applications ranging from biomedical to automobile, along with day to day applications according to the properties of the polymers [17]. For several areas, such as tissue engineering, orthopedic implants, or other biomedical applications, PCL has been found to be one of the most promising polymeric materials due to its lower melting temperature, easy processability, biodegradability, lower glass transition temperature, nontoxicity, relatively higher thermal degradation temperature, etc. [18]. However, it has been found that PCL has inherently unique characteristics, and certain limitations of this polymer restrict its direct use in the targeted applications [19]. The desired applications often require specific properties which are difficult to determine from a pristine polymer.

In order to serve their purpose, biodegradable polymers such as PCL can be used in combination with each other. Moreover, different techniques, such as blending of polymers [20], copolymerization [21], fabrication of biocomposite with biofillers [22,23], grafting [24], crosslinking [25], etc., can be used to tailor the properties of the PCL based final product. Fabrication of the composite with fillers via blending, in situ polymerization [26], and reinforcement [27], is a promising approach to tailor the final product's properties, such as mechanical strength, toughness, stiffness, biodegradability, and behavior of shape memory. Several organic and inorganic materials, such as carbon based fillers, chitosan [28], lignin [29], graphene [30], cellulose [31], magnetic nanoparticles [32], and metal nanoparticles [33] can be used as fillers for the development of biodegradable polyurethanes. Among these fillers, chitosan, which is often used as both filler and polymer matrix, is a promising biopolymer. Chitosan, which can be utilized in biomedical applications, is recognized as nontoxic, biodegradable, and antibacterial in nature. Chitosan can be produced using a process of chitin deacetylation, which is derived from the bioresource.

It is known that chitosan has a melting point near degradation temperature, which restricts its thermal processing both in its pristine form and with other polymers. The presence of free amino functional groups in the chitosan molecule aids in solubilizing it in acidic media by functional protonation. Additionally, it is challenging to obtain a uniform distribution of chitosan throughout the polymer matrix due to its hydrophilic nature. Several researchers have developed different process to modify chitosan and use it in biocomposite fabrication [34,35]. Various research groups are working worldwide to utilize chitosan in shape memory applications. By incorporating bioactive glass nanoparticles obtained from the sol-gel process, Leite et al. [36] have synthesized chitosan spheres. They reported a shape memory effect triggered by hydration and used for healing bones and controlling drug release. Bao et al. [37] have developed chitosan functionalized poly(lactic-co-glycolic acid) (PLGA) microspheres containing lysozyme. They have used high intensity focused ultrasound to stimulate the shape memory effect. Zhang et al. [38] used microwaves as a shape memory stimulant in PCL foam and found improvement in the recovery speed. Similarly, Shou et al. [39] have developed near infrared light (NIR) responsive shape memory films via photo-initiated polymerization of PCL with acryloyl terminal groups in the presence of gold nanorods. They conceptualized a controllable shape memory PCL, embedded with a gold nanorod film, that can actuate in a physiological temperature using NIR situated at a remote place. However, as per our knowledge, the use of chitosan as a filler in a PCL matrix by in situ reaction for a shape memory application has not been explored.

Therefore, the present work is focused on the development of a thermo-responsive shape memory biodegradable PCL based polyurethane and its composites. Herein, the chitosan flakes (nm to µm scale) react with the PCL diol, thereby providing a crosslinking point in the polymer matrix, which potentially enhances the shape memory effect along with the mechanical properties of the PCL based polyurethane. Several techniques, such as universal testing machine (UTM), thermogravimetric analysis (TGA), differential scanning calorimetry (DSC), fourier transform infrared (FTIR), and X-ray diffraction (XRD) have been used to characterize the prepared polyurethane biocomposites.

2. Materials and Methods

2.1. Materials

Chitosan was derived from shrimp shells with ≥75% deacetylation, which were obtained from Sigma Aldrich (Reykjavik, Iceland). PCL diol with an average molecular weight (Mn) of 2000 g/mole was purchased from Sigma Aldrich (Hong kong, China). 1,6-hexamethylene diisocynate (HMDI) was supplied by Daejung (Siheung, Korea). Dibutyltin dilaurate (DBDTL), used as catalyst, was provided by Sigma (St. Louis, MO, USA). Acetone and chloroform were procured from J.T. Baker Avantor (Center Valley, PA, USA). N,N-Dimethylformamide (DMF) and 1,4-butanediol (BDO) were purchased from Tedia Company Inc. (Fairfield, OH, USA) and Aldrich (Steinheim, Germany), respectively. Hydrogen peroxide (H_2O_2) was received from Samchun (Seoul, Korea). All the received chemicals were used without further purification, unless otherwise stated.

2.2. Preparation of Polyurethane Biocomposite

It is known that chitosan is available in the form of relatively large flakes and insoluble in DMF, due to strong glycosidic linkage in the backbone molecules. Initially, 10 g of chitosan was added to 100 mL of hydrogen peroxide (30%) by stirring (MS-DMS633, Mtops, Yangju, Korea) at 40 °C for 24 h to increase the dispersion and allow for swelling. After swelling, the prepared slurry was added to a sufficient amount of acetone to precipitate the chitosan. The swelling and precipitation processes are known to reduce the size of the obtained flakes and weaken glycosidic linkages. The obtained purified chitosan was washed with an adequate amount of deionized water several times until a neutral pH was obtained; this process was followed by freeze drying (Ilshin Lab Co. Ltd., Seoul, Korea) for 24 h. The resulting chitosan (the required amount) was allowed to swell in 10 mL of DMF for 24 h in order to use it for further biocomposite fabrication.

The Chitosan promoted PCL based shape memory polyurethane biocomposite was prepared according to the following steps. The required amount of PCL diol was dropped into a two-neck round-bottom flask equipped with a magnetic stirring bar and an inert environment kept at 40 °C on a stirrer with a hot mantle. Subsequently, the predetermined amount of chitosan swollen in DMF was added to the flask to maintain at 5%, 7.5%, 10%, and 20% against the PCL. After dissolution of the PCL, the temperature was raised to 80 °C, and stirring was maintained at 200 rpm. HMDI was added to the solution dropwise under inert conditions, along with a few drops of DBTDL as a catalyst. The reaction continued for 2 h to form a diisocynate terminated prepolymer. BDO was added to the reaction system as a chain extender, and stirring continued. The chain extension was followed by the addition of chloroform under a reflux condition to reduce the viscosity of the reaction system. The solution was then poured on a polypropylene tray. The solution tray was kept in a fume hood to allow the evaporation of the chloroform, followed by drying in a vacuum oven at 80 °C for 24 h. Additionally, PCL polyurethane without chitosan was prepared by following the same procedure as described above for the comparison. All samples was prepared by maintaining the NCO:OH final ratio as 1:1.1. The obtained film was stored in a refrigerator for further characterization. The polyurethane biocomposites with 5%, 7.5%, 10%, and 20% chitosan content were denoted as PCL-PU-M5, PCL-PU-M7.5, PCL-PU-M10, and PCL-PU-M20, respectively, while neat PCL polyurethane was denoted as PCL-PU.

2.3. Characterization

Fourier transform infrared (FTIR) spectroscopy spectra were obtained on a Nicolet IR 200 (Thermo Scientific, Waltham, MA, USA) instrument at room temperature. The spectra of the thin film were recorded against air as reference in the range between 4000 and 600 cm^{-1} after 64 scans with 4 cm^{-1} resolution.

Differential scanning calorimetry (DSC Q2000, TA Instruments, New Castle, DE, USA) was used to measure melting, crystallization, and glass transitions of the sample under an inert environment. The pre-weighed sample (5–10 mg) was heated from 25 °C to 150 °C with a rate of 10 °C/min and kept under isothermal conditions for 5 min to eliminate its thermal history. The sample was then cooled to 25 °C with a rate of 10 °C/min, kept isothermal for 5 min, and heated again to 150 °C with the same rate.

Thermogravimetric analysis (D-TGA, SDT 2960, TA Instruments) was carried out by heating 2–5 mg of the sample from room temperature to 500 °C with a rate of 10 °C/min under a N_2 environment.

X-ray diffraction (XRD) spectra were recorded using an automated multipurpose X-ray diffractometer (JP/SmartLab, 9kW, Rigaku, Tokyo, Japan) equipped with copper (Cu) Kα radiation (1.540593 Å) as an X-ray source (40 kV, 200 mA) with a monochromatic filter along with a detector (SC-70). The samples were analyzed in the range of 5–30° at a scan rate of 2°/min with a 0.02° resolution.

Contact angle measurement equipment (Model GSM) supplied by Surfacetech Co. Ltd. (Gwangju, Korea) was used to measure the contact angle of 2 μL of deionized water dropped on the surface of the prepared biocomposites. The films were pasted on a cleaned glass slide prior to measurement, and the angle was determined after 2 min of drop stabilization. The contact angles were determined from at least three different regions on each sample surface along with standard deviation.

Tensile strength and percent elongation of the prepared biocomposite samples were analyzed using a universal testing machine (UTM) (LRK-5kN, Cometech, QC-M2014, NTS Technology Co. Ltd. Chengdu, China). The specimens (5 mm width, 0.5–1.0 mm thickness, and 20 mm gauge length) were fixed and analyzed under a UTM equipped with a 5 kN load cell at a fixed cross-head speed of 5 mm/min in tensile mode. The data obtained were analyzed using AMIS 1.1.6 software and reported as the mean of at least five replicates of each sample along with standard deviation.

The topography of the fractured samples was examined using a field emission scanning electron microscope (FESEM). The samples were gold coated in a sputtering unit and characterized using FESEM (Ultra Plus, Zeiss GmbH) at an accelerating voltage of 3 kV after placing it on a carbon tape. The micrograph was analyzed using Gwyddion version 2.25 software (Cszech Metrology Institute, Brno, Cszech Republic).

The gel content, formed due to the formation of a cross-linked three-dimensional network, was quantified using Equation (1):

$$\text{Gel content}(\%) = \frac{W_{gel}}{W_i} \times 100 \quad (1)$$

where W_i and W_{gel} represent the initial weight of the prepared biocomposite and the weight of dried gel, respectively, after washing with chloroform and then vacuum drying.

The shape memory behavior of the prepared biocomposite was analyzed by performing a tensile test in warm water (50–60 °C) and hot oven (50 °C) conditions. The test was conducted in two environments: (1) Tensile pull and recovery in warm water, and (2) shape setting and recovery in a hot air condition. The sample was deformed by a tensile pull at 50 °C until a 100% strain was achieved. Next, the stretched shape was fixed by cooling the sample to 20 °C for 5 min, and the length was measured. The stretched and fixed sample was dropped in warm water (50–60 °C), and the length was measured after the recovery of the sample from the water. The shape fixing ratio and shape recovery ratio were calculated as per the following equations:

$$\text{Shape fixing ratio}\left(R_f\right) = \frac{(L_s - L_i)}{L_i} \times 100 \tag{2}$$

$$\text{Shape recovery ratio } (R_r) = \frac{\left(L_o - L_f\right)}{L_i} \times 100 \tag{3}$$

where L_o, L_s, L_f, and L_i represent the length of the sample after stress, the length of the sample after the release of stress, the length of the recovered sample, and the initial length of the sample, respectively.

3. Results and Discussion

3.1. Synthesis of the PCL Based Polyurethane Biocomposite

The molecular functional groups in the PCL-PU and PCL-PU-chitosan biocomposites were examined using FTIR. The spectra, shown in Figure 1, display the characteristic peaks for the PCL-PU and chitosan. The PCL-PU characteristic peaks at 1732 cm^{-1}, 2947 cm^{-1}, and 3332 cm^{-1} are assigned to the stretching vibration of the carbonyl group, the aliphatic C–H bond, and the N–H bond, respectively [40]. The band at 3442 cm^{-1} corresponds to the N–H and OH stretching vibration, whereas peaks at 2947 cm^{-1} and 2868 cm^{-1} are related, respectively, to the C–H symmetric and asymmetric stretching in chitosan. The bands around 1658 cm^{-1}, 1571 cm^{-1}, and 1326 cm^{-1} correspond to the C=O stretching of amide I, the N–H bending of amide II, and the C–N stretching of amide III of the chitosan acetyl group, respectively. The peaks merged between 1195 cm^{-1} and 936 cm^{-1} are related to the asymmetric stretching of the C–O–C backbone and the skeletal C–O stretching in pristine chitosan, respectively [41]. In the case of PCL-PU and other biocomposites, the peaks corresponding to the C–N stretching and N–H in-plane bending in the C–N–H functional group are located between 1643 cm^{-1} and 1504 cm^{-1}. A shoulder found on the absorption peak is attributed to the carbonyl group in PCL-PU. The peak at 1732–1724 cm^{-1} corresponds to the vibration of the amide group due to the urethane block and chitosan molecules. The stretching band at 3520 cm^{-1} corresponds to the hydroxyl functional group in PCL diol (not shown in the figure), which disappeared, while a new band, attributed to the N–H stretching, appears at 3332 cm^{-1}, thereby confirming the formation of polyurethane. The presence of peaks related to PCL and chitosan in the prepared biocomposites confirms the formation of polyurethane and chitosan biocomposites.

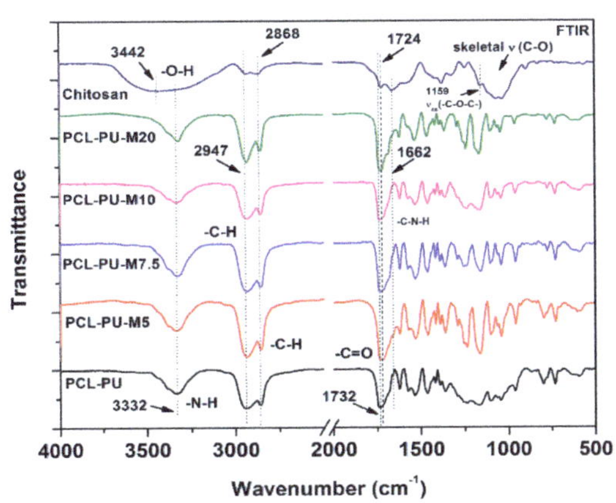

Figure 1. FTIR spectra of Chitosan, polycaprolactone-polyurethane (PCL-PU), and PCL-PU-Chitosan biocomposites.

3.2. Thermal Transitions of Prepared Polyurethane Biocomposites

The prepared PCL based polyurethane biocomposites were analyzed using DSC, and the thermograms are shown in Figure 2. The heating cycle (Figure 2a) of the biocomposite shows the presence of PCL in crystalline form, with a melting temperature around 40 °C. The melting temperature of biocomposites is found to be the same with PCL-PU, whereas the enthalpy of melting is reduced from 23.7 J/g to 11.0 J/g as the content of chitosan in the polymer matrix is increased. The reduction in the melting enthalpy suggests a reduction in the crystallinity of the PCL with increasing chitosan content. The presence of chitosan possibly hinders PCL chain mobility, thereby restricting the orientation of chains and forming a crystalline domain, which may be responsible for the reduction in the crystallinity of the biocomposite. Chitosan nano-sized flakes are connected to PCL chains, which form the crosslinking point in the polyurethane system, resulting in an increased amorphous domain. In the case of 20% chitosan content, the melting temperature of the biocomposite is higher (46.7 °C), which could be attributed to the phase separation of the PCL and chitosan molecules, and the crystalline domain is reduced to 11.0 J/g. The exothermic peak of the biocomposite in the cooling cycle is found to be shifted towards a lower temperature, as presented in Figure 2b. The reduction in the melt crystallization temperature from 2.0 °C for PCL-PU to −9.6 °C for PCL-PU-M10 suggests that the PCL chains in biocomposites need less energy to be crystallized. However the enthalpy of crystallization is reduced from 22.6 J/g to 8.3 J/g, which could be due to the anti-nucleation effect of the chitosan flakes.

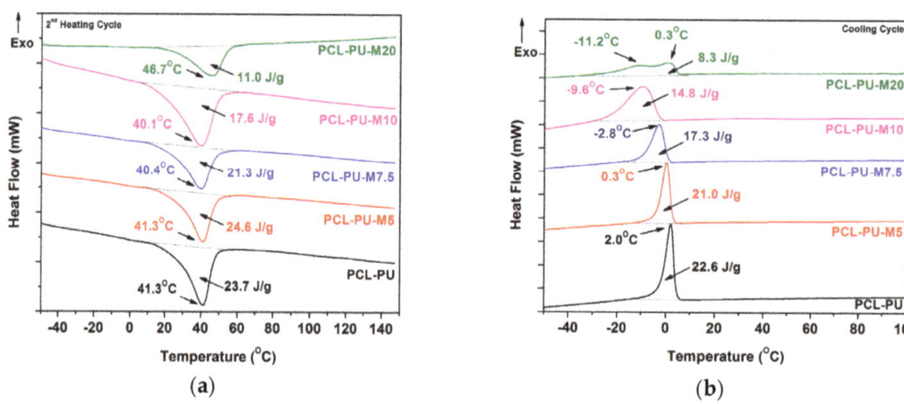

Figure 2. Differential scanning calorimetry (DSC) thermograms: (**a**) Second heating and (**b**) cooling cycle of the PCL-PU and PCL-PU-Chitosan biocomposites.

3.3. X-ray and Contact Angle Measurements

The biocomposite was characterized using X-ray diffraction spectroscopy, as shown in the Figure 3a. The characteristic peaks for PCL at 21.3°and 23.6° correspond to the (110) and (200) planes of the PCL crystal, whereas the peak at 19.3° is related to the (110) plane of the chitosan micro crystalline domain reflection [42,43]. The peak corresponding to chitosan has a significantly lower intensity in PCL-PU polyurethane biocomposites, whereas it is visible in the biocomposite with 20% chitosan content, indicating the presence of phase separated domains. No peak shifting was observed, indicating no change in the crystal structure of the PCL and chitosan.

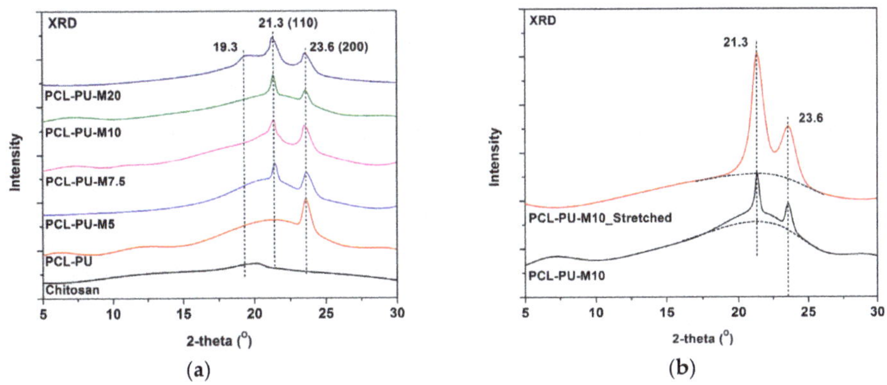

Figure 3. (**a**) X-ray diffraction of the PCL-PU and PCL-PU-Chitosan biocomposite and (**b**) X-ray diffraction comparison of the stretched and non-stretched biocomposite sample.

Contact angle measurement of the prepared biocomposites was employed to understand their interaction with water molecules. The data are presented in Figure 4, and the contact angle was found to increase from around 69° to ~95°, alongside the increase in the content of chitosan from 0% to 10%. It is known that wettability is based on the chemical composition, surface topography, and surface free energy of the substance. As the chitosan flake content increases in the biocomposite, the contact angle gradually increases, possibly due to the reduction in surface free energy. The presence of chitosan flakes also affects the surface roughness, which may lead to an increase in the water contact angle. It is observed that the contact angle reduces to ~80° in a case of 20% chitosan, suggesting phase separation and a reduction in hydrophobicity.

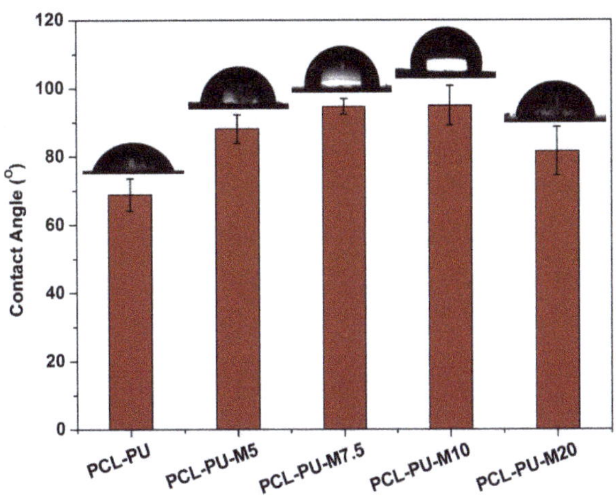

Figure 4. Contact angle measurement of PCL-PU and PCL-PU-Chitosan biocomposites.

3.4. *Mechanical Properties of Polyurethane Biocomposites*

The presence of nano-sized chitosan flakes influences the mechanical properties of the prepared polyurethane biocomposites. It is found that chitosan nanoflakes provide crosslink sites, thereby providing strength and imparting shape memory properties to the polyurethane biocomposite. The representative load-elongation curve and ultimate tensile strength (UTS), along with the elongation at break for PCL-PU and PCL-PU-Chitosan polyurethane biocomposites, are shown in Figure 5a,b.

The UTS for PCL-PU was found to be 5.2 MPa, which increases by approximately 130% and reaches 12 MPa for the PCL-PU-M10 composite. The strain hardening due to the strong interfacial bonding between the chitosan and PCL chains and the presence of chitosan nano-sized flakes possibly resist the mechanical pull after a certain elongation, ultimately resulting in increased UTS. The elongation at break for PCL-PU was found to be 102%, which increases to 598% after the addition of 10% chitosan in the polyurethane biocomposite, indicating a significant improvement in the ductility of the PCL-PU biocomposite with integration of chitosan flakes. The increase in elongation could be the result of the uncoiling of coiled PCL chains present in the polyurethane biocomposite. It is also known that enhancement in the elongation at break or ductility delays the fracture, which reduces the risk of abrupt mechanical failure of the polyurethane biocomposite. As reported [44], the use of chitosan as a filler may improve the UTS of the polymer matrix with compromising elongation at break, whereas an increase in tensile strength along with elongation at break is observed in the present study. Increased elongation at break could be due to the chemical interaction of chitosan nanoflakes with PCL chains, which increase the coiling of the PCL chains at particular crosslink sites. The integration of chitosan with PCL chains can act as a bridge between the uncoiling and sliding chains, thereby lengthening the breakage process. However, in the case of PCL-PU-M20, the elongation at break is reduced to 5.8%. The increased content of chitosan (20%) in the PCL based polyurethane matrix ensures an improvement of ~21% in tensile strength compared to PCL-PU. However, the presence of chitosan in higher amounts may induce voids during the elongation process, ultimately resulting in breakage and a rupture of bonds leading to a reduction in the elongation at break.

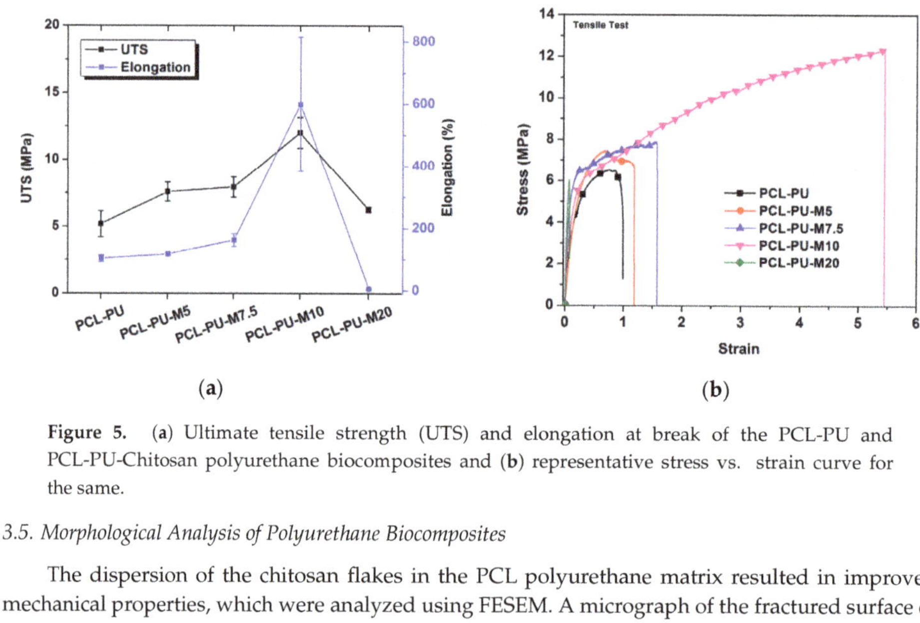

Figure 5. (**a**) Ultimate tensile strength (UTS) and elongation at break of the PCL-PU and PCL-PU-Chitosan polyurethane biocomposites and (**b**) representative stress vs. strain curve for the same.

3.5. Morphological Analysis of Polyurethane Biocomposites

The dispersion of the chitosan flakes in the PCL polyurethane matrix resulted in improved mechanical properties, which were analyzed using FESEM. A micrograph of the fractured surface of the PCL-PU and PCL-PU-Chitosan biocomposites, which displays a homogeneous dispersion of the chitosan flakes, is shown in Figure 6. It can be seen from the figure that the chitosan flakes reduced in size during biocomposite fabrication and were found to be around ~150 nm in width and ~600–1000 nm in length, whereas the PCL-PU has a relatively smooth surface.

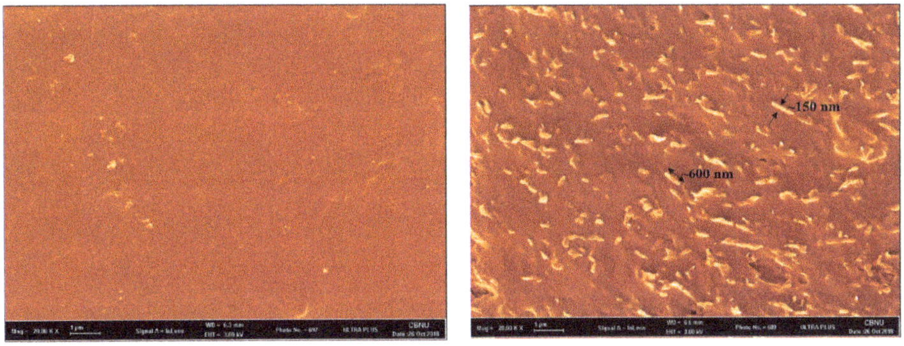

(**a**) (**b**)

Figure 6. Representative field emission scanning electron microscope (FESEM) micrographs of (**a**) the PCL-PU and (**b**) the PCL-PU-Chitosan biocomposite (Scale bar: 1 μm).

3.6. Thermal Degradation of Polyurethane Biocomposites

The thermal behavior of the prepared polyurethane biocomposites was analyzed using TGA and presented in Figure 7. The onset degradation temperature of the chitosan was found to be 135 °C, whereas the maximum degradation rate was observed around 235 °C. It can be surmised that the presence of chitosan may affect the degradation behavior of the biocomposite due to its relatively lower degradation temperature. Conversely, no adverse effect was found in the biocomposite due to the presence of chitosan. The data for the degradation temperature and residue content is listed in Table 1. The onset degradation temperature of the polyurethane biocomposite increased from 310 °C to 346 °C, alongside the increase in the content of chitosan from 5% to 10%. Similarly, the maximum degradation temperature improved from 360 °C to 384 °C, with an upsurge in the content of chitosan from 0% to 10%. An increase in the content of chitosan to 20% lead to a reduction in onset degradation (327 °C) as well as maximum degradation temperature (379 °C). The effect of chitosan in the present polyurethane biocomposite can be better understood by estimating the degradation temperature at 10% weight loss (T_{10}). In Table 1, the T_{10} for PCL-PU is 309 °C, which reduced to 276 °C after the addition of 5–10% chitosan, and again increased to 318 °C. This change could be due to the increased interaction of chitosan and PCL-diol, which stabilizes the polymer system by means of increased cross-linking points. A reduction to 255 °C was found in a case of 20% chitosan content, suggesting phase separation. The residual content after thermal degradation, which is a nondegradable carbon residue of chitosan, was also found to increase with an increase in the content of chitosan. Furthermore, it can be derived from the TGA analysis that the presence of chitosan, which has a lower degradation temperature, had no adverse effect on the thermal stability of the polyurethane biocomposite. The crosslinking between the PCL diol and chitosan molecule could be responsible for the stabilization of the polymer system. The crosslinking between the PCL diol and chitosan was also confirmed by the gel content analysis.

Table 1. Thermal transitions and percentage residue for PCL-PU and PCL-PU-Chitosan biocomposites.

Sample Name	T_{onset} (°C)	T_{max} (°C)	T_{10} (°C)	T_{50} (°C)	Residue (%)
Chitosan	135	235	135	314	33.2
PCL-PU	315	360	309	357	0.0
PCL-PU-M5	310	364	276	354	2.6
PCL-PU-M7.5	311	366	276	352	4.7
PCL-PU-M10	346	384	318	381	5.8
PCL-PU-M20	327	379	255	379	10.8

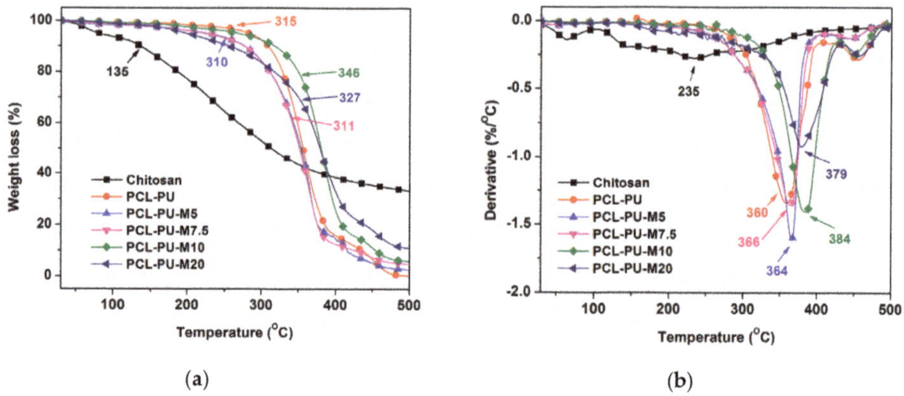

Figure 7. (**a**) Weight loss and (**b**) derivative weight loss of PCL-PU and PCL-PU-Chitosan biocomposite against temperature.

3.7. Gel Content of Polyurethane Biocomposites

The samples were suspended in chloroform in order to dissolve the non-crosslinked content, and the insoluble content was measured. A gel content of ~25.4%, which increased to 81.2% with an increase in the content of chitosan from 0% to 20% (Figure 8), was found for the PCL-PU. This improvement in the gel content suggests an enhancement in cross-linking that reduced the solubility of the polymer system. The presence of chitosan provided the cross-linking points which led to an increase in the gel content.

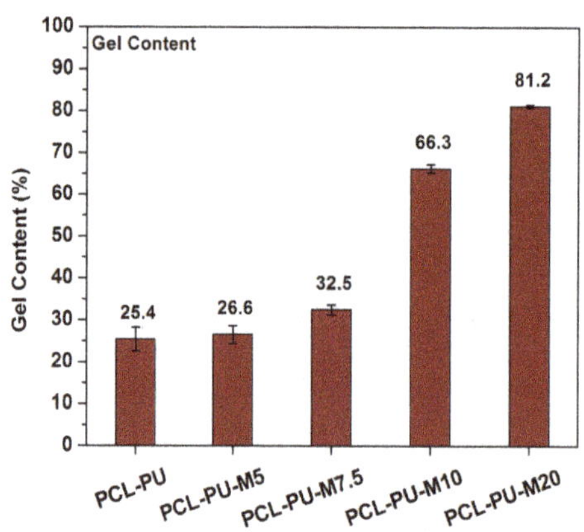

Figure 8. Gel content in the PCL-PU and PCL-PU-Chitosan biocomposites in chloroform.

3.8. Shape Memory Effect of PCL Based Polyurethane Biocomposites

Figure 9 demonstrates the shape memory effect of the prepared polyurethane biocomposite in a dry condition where the polyurethanes were initially fixed by heating them to 50 °C followed by cooling them to room temperature. It was found that the time required for shape recovery after the fixing was quite similar for the PCL-PU and PCL-PU-Chitosan biocomposites. The shape recovery of the biocomposite was also demonstrated in a wet condition, which is shown in Figure 10. A random shaped polymer sheet was punched with circular shapes and mechanically stretched at 50 °C followed by cooling to room temperature. The mechanical stretching led to a permanent deformation of the circled shapes, which were recovered by approximately 100% by placing the sample in warm water at

50 °C. For the quantitative analysis of the shape memory behavior of the polyurethane biocomposite, the shape fixing ratio and shape recovery ratio was measured and is presented in Table 2. It is seen that the shape fixing ratio improved from ~53% to approximately 100% after the incorporation of 10% chitosan. Similarly, the shape recovery ratio was found to be improved from ~88% to ~100%.

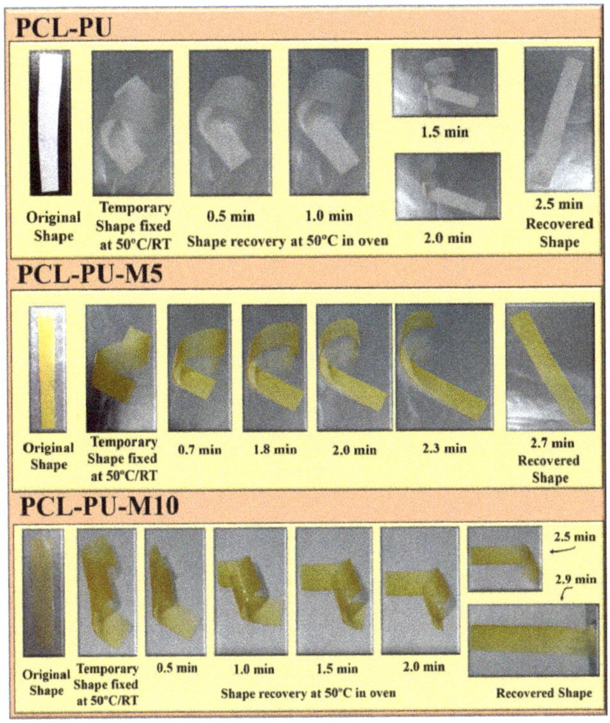

Figure 9. Demonstration of the shape memory effect in the PCL-PU-Chitosan biocomposite in a dry condition.

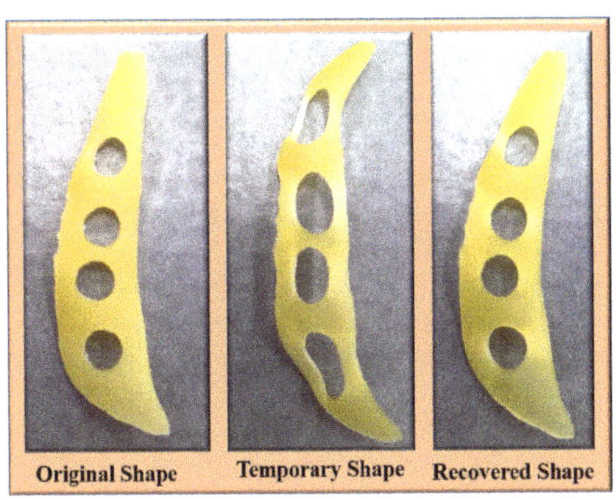

Figure 10. The shape memory ability of the prepared random shaped chitosan polyurethane composite for grip application (Temporary shape at room temperature using tensile force; recovered shape in water (wet condition) at 50 °C within 5 s).

Table 2. Shape fixing and shape recovery ratios of the PCL-PU and PCL-PU-Chitosan polyurethane biocomposites.

Polyurethane Samples	Shape Fixing Ratio (%)	Shape Recovery Ratio (%)
PCL-PU	53.1 ± 6.2	88.6 ± 1.3
PCL-PU-M5	81.2 ± 2.6	90.8 ± 1.2
PCL-PU-M7.5	93.4 ± 5.0	98.5 ± 0.5
PCL-PU-M10	99.5 ± 0.4	99.3 ± 0.7
PCL-PU-M20	Not stretchable	Not stretchable

The mechanical analysis revealed that the toughness of the materials was drastically improved, which could be the result of the increased crosslinked nodes in the polymer system. The shape recovery of the prepared material from an elongated (~500% strain) or permanently deformed condition is shown in Figure 11.

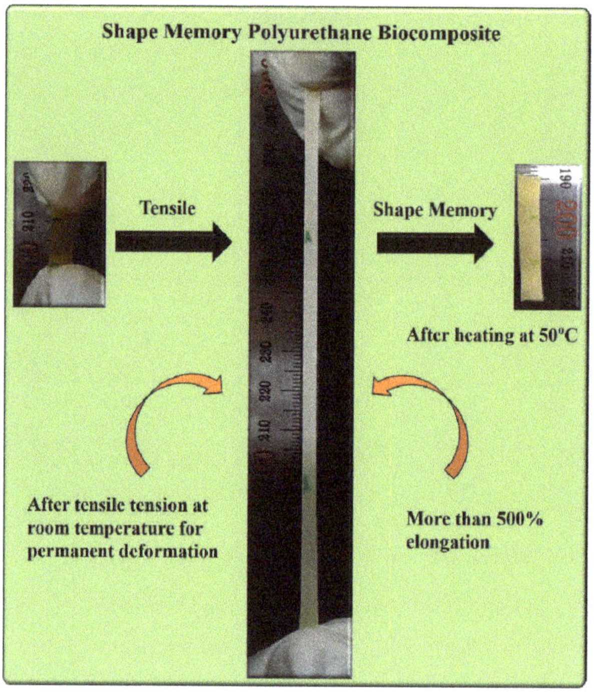

Figure 11. Shape memory behavior of the polyurethane chitosan biocomposite after tensile tension.

The coiling and uncoiling of the PCL chains could be responsible for the shape memory ability, which greatly corresponds to the crosslink content in the polymer matrix. It was found that the polyurethane biocomposite with chitosan retains its temporary shape even after the removal of stress and sustained restraint. The developed restraint in the shape memory polyurethane biocomposite due to tensile pull is possibly released during thermal softening around the melting of the PCL crystallites. An upsurge in the temperature triggers the melting of PCL crystallites which are formed due to a stress induced crystallization phenomenon during tensile pull, and the material tends to recover its original shape. The chitosan nanoflakes work as rigid crosslinking nodes covalently bonded between extendable PCL chains. These cross-linking nodes could be responsible for putting back polymer chains in response to an external stimulus. On the other hand, the formation of crystallites facilitates a proficient and quick formation of the switching domain, as observed in the DSC analysis (the melting of PCL crystallites in the range of 40–50 °C) and also confirmed from the X-ray analysis.

XRD analysis of the stretched and non-stretched (PCL-PU-chitosan) samples is shown in Figure 3b. It can be seen that the intensity of the characteristic peaks for PCL at 21.3 ° and 23.6 ° is enhanced, suggesting that the crystallinity of the biocomposite is increased. The enhancement in the crystallinity could be the consequence of stress induced crystallization during tensile work. The uncoiling of coiled PCL chains, and arrangement of the same, increases the crystallinity of the biocomposite and possibly enhances the shape memory even after permanent deformation. Also, the shape recovery phenomenon of PCL based chitosan polyurethane biocomposite is demonstrated in Video S1.

4. Conclusions

We have successfully demonstrated the fabrication of a PCL based shape memory polyurethane biocomposite via in situ incorporation of chitosan. The developed biocomposite with 10% chitosan content had ~130% enhancement in ultimate tensile strength and more than 500% elongation at break in comparison to pristine PCL-PU. The electron microscopic analysis confirmed the homogeneous distribution of chitosan flakes with ~150 nm width in the polymer matrix. X-ray and DSC analysis confirmed the presence of PCL crystallites, which may act as a switching point for the shape memory behavior. X-ray diffraction analysis also confirmed the development of crystallites, resulting from a stress induced crystallization process during tensile pull, which may retain the shape and melting of the crystallites, thereby stimulating shape recovery (with ~100% shape recovery ratio), even after permanent deformation. The biodegradable polyurethane biocomposite also demonstrates relatively higher thermal stability (T_{max} at ~360 °C). Overall, the fabricated shape memory polyurethane biocomposite possesses a unique shape memory behavior, even after permanent deformation to more than a 500% strain, which has extreme potential in several biomedical applications.

Supplementary Materials: The following are available online at http://www.mdpi.com/2079-4991/9/2/225/s1, Video S1: Video demonstrating the shape recovery phenomenon of PCL based chitosan polyurethane biocomposite.

Author Contributions: A.G. developed and designed the research, executed experiments, analyzed data and wrote the manuscript. B.S.K. designed the research and discussed the results. All authors reviewed the manuscript.

Funding: This work was supported by Post-Doctoral Fellowship Program funded by the Ministry of Education of the Republic of Korea through the Chungbuk National University in 2018–2019 and the National Research Foundation of Korea (NRF-2017R1A2B4002371).

Conflicts of Interest: The authors declare no conflict of interest.

References

1. Wadood, A. Brief overview on nitinol as biomaterial. *Adv. Mat. Sci. Eng.* **2016**, *2016*, 9. [CrossRef]
2. Ortega, A.M.; Yakacki, C.M.; Dixon, S.A.; Likos, R.; Greenberg, A.R.; Gall, K. Effect of crosslinking and long-term storage on the shape-memory behavior of (meth)acrylate-based shape-memory polymers. *Soft Matter* **2012**, *8*, 7381–7392. [CrossRef]
3. Xiao, X.; Kong, D.; Qiu, X.; Zhang, W.; Liu, Y.; Zhang, S.; Zhang, F.; Hu, Y.; Leng, J. Shape memory polymers with high and low temperature resistant properties. *Sci. Rep.* **2015**, *5*, 14137. [CrossRef] [PubMed]
4. Memarian, F.; Fereidoon, A.; Ghorbanzadeh Ahangari, M. Effect of acrylonitrile butadiene styrene on the shape memory, mechanical, and thermal properties of thermoplastic polyurethane. *J. Vinyl Addit. Techn.* **2018**, *24*, E96–E104. [CrossRef]
5. Sakurai, K.; Tanaka, H.; Ogawa, N.; Takahashi, T. Shape-memorizable styrene-butadiene block copolymer. I. Thermal and mechanical behaviors and structural change with deformation. *J. Macromol. Sci. B* **1997**, *36*, 703–716. [CrossRef]
6. Hu, J.; Yang, Z.; Yeung, L.; Ji, F.; Liu, Y. Crosslinked polyurethanes with shape memory properties. *Polym. Int.* **2005**, *54*, 854–859. [CrossRef]

7. Li, P.; Han, Y.; Wang, W.; Liu, Y.; Jin, P.; Leng, J. Novel programmable shape memory polystyrene film: A thermally induced beam-power splitter. *Sci. Rep.* **2017**, *7*, 44333. [CrossRef] [PubMed]
8. Xie, T. Recent advances in polymer shape memory. *Polymer* **2011**, *52*, 4985–5000. [CrossRef]
9. Behl, M.; Lendlein, A. Shape-memory polymers. *Mater. Today* **2007**, *10*, 20–28. [CrossRef]
10. Seymour, R.B.; Kauffman, G.B. Polyurethanes: A class of modern versatile materials. *J. Chem. Edu.* **1992**, *69*, 909. [CrossRef]
11. Akindoyo, J.O.; Beg, M.D.H.; Ghazali, S.; Islam, M.R.; Jeyaratnam, N.; Yuvaraj, A.R. Polyurethane types, synthesis and applications—A review. *RSC Adv.* **2016**, *6*, 114453–114482. [CrossRef]
12. Puszka, A.; Kultys, A. New thermoplastic polyurethane elastomers based on aliphatic diisocyanate. *J. Therm. Anal. Calorim.* **2017**, *128*, 407–416. [CrossRef]
13. Mi, H.-Y.; Jing, X.; Napiwocki, B.N.; Hagerty, B.S.; Chen, G.; Turng, L.-S. Biocompatible, degradable thermoplastic polyurethane based on polycaprolactone-block-polytetrahydrofuran-block-polycaprolactone copolymers for soft tissue engineering. *J. Mat. Chem. B* **2017**, *5*, 4137–4151. [CrossRef] [PubMed]
14. Ping, P.; Wang, W.; Chen, X.; Jing, X. Poly(ε-caprolactone) polyurethane and its shape-memory property. *Biomacromolecules* **2005**, *6*, 587–592. [CrossRef]
15. Lendlein, A.; Langer, R. Biodegradable, elastic shape-memory polymers for potential biomedical applications. *Science* **2002**, *296*, 1673–1676. [CrossRef] [PubMed]
16. The future of plastic. *Nat. Commun.* **2018**, *9*, 2157. [CrossRef] [PubMed]
17. Gross, R.A.; Kalra, B. Biodegradable polymers for the environment. *Science* **2002**, *297*, 803–807. [CrossRef] [PubMed]
18. Malikmammadov, E.; Tanir, T.E.; Kiziltay, A.; Hasirci, V.; Hasirci, N. PCL and PCL-based materials in biomedical applications. *J. Biomat. Sci. Polym. Edit.* **2018**, *29*, 863–893. [CrossRef] [PubMed]
19. Gupta, A.; Katiyar, V. Cellulose functionalized high molecular weight stereocomplex polylactic acid biocomposite films with improved gas barrier, thermomechanical properties. *ACS Sustain. Chem. Eng.* **2017**, *5*, 6835–6844. [CrossRef]
20. Pack, S.; Lewin, M.; Rafailovich, M.H. A review of engineering biodegradable polymer blends: Morphology, mechanical property, and flame retardancy. In *Fire and Polymers VI: New advances in Flame Retardant Chemistry and Science*; American Chemical Society: Washington, DC, USA, 2012; Volume 1118, pp. 427–443.
21. Yılmaz, M.; Eğri, S.; Yıldız, N.; Çalımlı, A.; Pişkin, E. Ring-opening copolymerization of l-lactide and ε-caprolactone in supercritical carbon dioxide using triblock oligomers of caprolactone and PEG as stabilizers. *Polym. J.* **2011**, *43*, 785. [CrossRef]
22. Gupta, A.; Prasad, A.; Mulchandani, N.; Shah, M.; Ravi Sankar, M.; Kumar, S.; Katiyar, V. Multifunctional nanohydroxyapatite-promoted toughened high-molecular-weight stereocomplex poly(lactic acid)-based bionanocomposite for both 3D-printed orthopedic implants and high-temperature engineering applications. *ACS Omega* **2017**, *2*, 4039–4052. [CrossRef] [PubMed]
23. Gupta, A.; Pal, A.K.; Woo, E.M.; Katiyar, V. Effects of amphiphilic chitosan on stereocomplexation and properties of poly(lactic acid) nano-biocomposite. *Sci. Rep.* **2018**, *8*, 4351. [CrossRef] [PubMed]
24. Flesch, C.; Delaite, C.; Dumas, P.; Bourgeat-Lami, E.; Duguet, E. Grafting of poly(ε-caprolactone) onto maghemite nanoparticles. *J. Polym. Sci. A Polym. Chem.* **2004**, *42*, 6011–6020. [CrossRef]
25. Peponi, L.; Navarro-Baena, I.; Sonseca, A.; Gimenez, E.; Marcos-Fernandez, A.; Kenny, J.M. Synthesis and characterization of PCL-PLLA polyurethane with shape memory behavior. *Eur. Polym. J.* **2013**, *49*, 893–903. [CrossRef]
26. Lopez-Figueras, L.; Navascues, N.; Irusta, S. Polycaprolactone/mesoporous silica mcm-41 composites prepared by in situ polymerization. *Particuology* **2017**, *30*, 135–143. [CrossRef]
27. Ahmad, A.F.; Abbas, Z.; Aziz, S.A.; Obaiys, S.J.; Zainuddin, M.F. Synthesis and characterisation of nickel oxide reinforced with polycaprolactone composite for dielectric applications by controlling nickel oxide as a filler. *Results Phys.* **2018**, *11*, 427–435. [CrossRef]
28. Sarasam, A.; Madihally, S.V. Characterization of chitosan–polycaprolactone blends for tissue engineering applications. *Biomaterials* **2005**, *26*, 5500–5508. [CrossRef] [PubMed]
29. Grząbka-Zasadzińska, A.; Klapiszewski, Ł.; Borysiak, S.; Jesionowski, T. Thermal and mechanical properties of silica–lignin/polylactide composites subjected to biodegradation. *Materials* **2018**, *11*, 2257. [CrossRef]
30. Duan, T.; Lv, Y.; Xu, H.; Jin, J.; Wang, Z. Structural effects of residual groups of graphene oxide on poly(ε-caprolactone)/graphene oxide nanocomposite. *Crystals* **2018**, *8*, 270. [CrossRef]

31. Benkaddour, A.; Jradi, K.; Robert, S.; Daneault, C. Grafting of polycaprolactone on oxidized nanocelluloses by click chemistry. *Nanomaterials* **2013**, *3*, 141–157. [CrossRef]
32. Thompson, Z.; Rahman, S.; Yarmolenko, S.; Sankar, J.; Kumar, D.; Bhattarai, N. Fabrication and characterization of magnesium ferrite-based PCL/Aloe vera nanofibers. *Materials* **2017**, *10*, 937. [CrossRef]
33. Xu, T.; Jin, W.; Wang, Z.; Cheng, H.; Huang, X.; Guo, X.; Ying, Y.; Wu, Y.; Wang, F.; Wen, Y.; et al. Electrospun CuO-nanoparticles-modified polycaprolactone @polypyrrole fibers: An application to sensing glucose in saliva. *Nanomaterials* **2018**, *8*, 133.
34. Gupta, A.; Mulchandani, N.; Shah, M.; Kumar, S.; Katiyar, V. Functionalized chitosan mediated stereocomplexation of poly(lactic acid): Influence on crystallization, oxygen permeability, wettability and biocompatibility behavior. *Polymer* **2018**, *142*, 196–208. [CrossRef]
35. Mourya, V.K.; Inamdar, N.N. Chitosan-modifications and applications: Opportunities galore. *React. Funct. Polym.* **2008**, *68*, 1013–1051. [CrossRef]
36. Leite, Á.J.; Caridade, S.G.; Mano, J.F. Synthesis and characterization of bioactive biodegradable chitosan composite spheres with shape memory capability. *J. Non-Cryst. Solids* **2016**, *432*, 158–166. [CrossRef]
37. Bao, M.; Zhou, Q.; Dong, W.; Lou, X.; Zhang, Y. Ultrasound-modulated shape memory and payload release effects in a biodegradable cylindrical rod made of chitosan-functionalized PLGA microspheres. *Biomacromolecules* **2013**, *14*, 1971–1979. [CrossRef] [PubMed]
38. Zhang, F.; Zhou, T.; Liu, Y.; Leng, J. Microwave synthesis and actuation of shape memory polycaprolactone foams with high speed. *Sci. Rep.* **2015**, *5*, 11152. [CrossRef]
39. Shou, Q.; Uto, K.; Iwanaga, M.; Ebara, M.; Aoyagi, T. Near-infrared light-responsive shape-memory poly(ε-caprolactone) films that actuate in physiological temperature range. *Polym. J.* **2014**, *46*, 492. [CrossRef]
40. Barikani, M.; Honarkar, H.; Barikani, M. Synthesis and characterization of polyurethane elastomers based on chitosan and poly(ε-caprolactone). *J. Appl. Polym. Sci.* **2009**, *112*, 3157–3165. [CrossRef]
41. Van de Velde, K.; Kiekens, P. Structure analysis and degree of substitution of chitin, chitosan and dibutyrylchitin by FT-IR spectroscopy and solid state ^{13}C NMR. *Carbohyd. Polym.* **2004**, *58*, 409–416. [CrossRef]
42. Baji, A.; Wong, S.-C.; Liu, T.; Li, T.; Srivatsan, T.S. Morphological and X-ray diffraction studies of crystalline hydroxyapatite-reinforced polycaprolactone. *J. Biomed. Mater. Res. B Appl. Biomater.* **2007**, *81B*, 343–350. [CrossRef] [PubMed]
43. Yuan, Y.; Chesnutt, B.M.; Haggard, W.O.; Bumgardner, J.D. Deacetylation of chitosan: Material characterization and in vitro evaluation via albumin adsorption and pre-osteoblastic cell cultures. *Materials* **2011**, *4*, 1399. [CrossRef] [PubMed]
44. Wan, Y.; Huang, J.; Zhang, J.; Yin, D.; Zheng, Z.; Liao, C.; Sun, S. Investigation of mechanical properties and degradability of multi-channel chitosan–polycaprolactone/collagen conduits. *Polym. Degrad. Stabil.* **2013**, *98*, 122–132. [CrossRef]

nanomaterials

MDPI

Article

Differential Cytotoxicity of Different Sizes of Graphene Oxide Nanoparticles in Leydig (TM3) and Sertoli (TM4) Cells

Sangiliyandi Gurunathan, Min-Hee Kang, Muniyandi Jeyaraj and Jin-Hoi Kim *

Department of Stem Cell and Regenerative Biotechnology, Konkuk University, Seoul 05029, Korea; gsangiliyandi@yahoo.com (S.G.); pocachippo@gmail.com (M.-H.K.); muniyandij@yahoo.com (M.J.)
* Correspondence: jhkim541@konkuk.ac.kr; Tel.: +82-2-450-3687

Received: 21 November 2018; Accepted: 18 January 2019; Published: 22 January 2019

Abstract: Graphene oxide (GO) is an common nanomaterial and has attracted unlimited interest in academia and industry due to its physical, chemical, and biological properties, as well as for its tremendous potential in applications in various fields, including nanomedicine. Whereas studies have evaluated the size-dependent cytotoxicity of GO in cancer cells, there have been no studies on the biological behavior of ultra-small graphene nanosheets in germ cells. To investigate, for the first time, the cyto- and geno- toxic effects of different sizes of GO in two different cell types, Leydig (TM3) and Sertoli (TM4) cells, we synthesized different sized GO nanosheets with an average size of 100 and 20 nm by a modification of Hummers' method, and characterized them by various analytical techniques. Cell viability and proliferation assays showed significant size- and dose-dependent toxicity with GO-20 and GO-100. Interestingly, GO-20 induced significant loss of cell viability and cell proliferation, higher levels of leakage of lactate dehydrogenase (LDH) and reactive oxygen species (ROS) generation compared to GO-100. Both GO-100 and GO-20 induced significant loss of mitochondrial membrane potential (MMP) in TM3 and TM4 cells, which is a critical factor for ROS generation. Furthermore, GO-100 and GO-20 caused oxidative damage to DNA by increasing the levels of 8-oxo-dG, which is formed by direct attack of ROS on DNA; GO-100 and GO-20 upregulate various genes responsible for DNA damage and apoptosis. We found that phosphorylation levels of EGFR/AKT signaling molecules, which are related to cell survival and apoptosis, were significantly altered after GO-100 and GO-20 exposure. Our results showed that GO-20 has more potent toxic effects than GO-100, and that the loss of MMP and apoptosis are the main toxicity responses to GO-100 and GO-20 treatments, which likely occur due to EGFR/AKT pathway regulation. Collectively, our results suggest that both GO-100 and GO-20 exhibit size-dependent germ cell toxicity in male somatic cells, particularly TM3 cells, which seem to be more sensitive compared to TM4, which strongly suggests that applications of GO in commercial products must be carefully evaluated.

Keywords: graphene oxide; Sertoli cells; Leydig cells; apoptosis; oxidative stress; mitochondrial membrane potential; DNA damage

1. Introduction

Carbon nanomaterials, such as fullerene, carbon nanotubes (CNTs), and graphene family materials, including few-layer graphene, reduced graphene oxide (GO), graphene nanosheets, and GO, have attracted much attention from the scientific community, as well as industry, due to their wide applications in areas of biotechnology, biomedical engineering, and nanodevices. Among several carbon nanomaterials, graphene nanoparticles exhibit extensive usage in a variety of applications due to high surface-to-volume ratio, high mechanical strength, flexible nature, ease of functionalization, and biocompatibility [1–4]. Compared with other carbon nanomaterials, GO provides a larger surface area and has better solubility, which leads to it being a suitable candidate for various biomedical applications, compared with other carbon-based materials.

Graphene oxide is a two-dimensional nanomaterial with one-atom thick planar sheets of sp^2-bonded carbon atoms, having many oxygenated functional groups. Graphene can be synthesized using various methods, physical mechanical cleavage, the 'Scotch tape' method [5], arc discharge [6], chemical methods [7], chemical vapor deposition [7], chemical oxidation [8] or longitudinal unzipping of carbon nanotubes [9]. The method of synthesis determines the size, morphology, solubility, toxicity, and biocompatibility of graphene. Zhang et al. (2013) synthesized uniform different sizes graphene oxide nanosheets with an average size of 50 nm at high yield, and also synthesized graphene sheets which exhibited fluorescent properties, outstanding stability, and excellent biocompatibility with lower cytotoxicity and higher cellular uptake. Generally, uniform-sized nanomaterials with diameters of less than 100 nm are more suitable for intracellular applications such as imaging and drug delivery [10].

The evaluation of in vitro cytotoxicity is a crucial step for the development of nanoparticle-based formulations for nanomedicine, biomedical engineering, and biotechnological applications. Toxicity or biocompatibility depends on size, morphology, dose, time of incubation, and type of cells involved. The morphology of GO nanomaterials is crucial to their precise interactions and functioning in cells. For example, Zhang et al. [11] found that the toxicity of GO in rat pheochromocytoma PC12 cells is dose-dependent, and that the size of the particles ranged from 10–100 nm, with thicknesses of 3–5 nm. Toxicity was observed at high concentrations of 100 μg/mL, and no toxicity was observed at 0.01–10 μg/mL. Akhavan et al. [12] reported size-dependent toxicity in human mesenchymal stem cells (hMSCs). They observed significant size-dependent cytotoxicity with rGONPs (11 ± 4 nm) based on lateral size dimensions (3.8 ± 0.4 μm). Chang et al. [13] observed size-dependent toxicity in A549 (human lung adenocarcinoma) cells. These results suggest that small GO sheets (160 ± 90 nm) showed significant toxicity compared to larger GO sheets (430 ± 300 nm and 780 ± 410 nm). Chowdhury et al. (2014) [14] prepared a variety of sizes using bath and probe sonication, and the effect was investigated in A549 and human breast cancer cells (MCF-7). The authors found that graphene nanoribbons (GNRs) subjected to 1 minute of probe sonication can result in significant decreases in the overall metabolic state of cells in vitro.

An in vitro study revealed that oxidized graphene was the most cytotoxic, inducing mitochondrial and plasma-membrane damage, and suggesting low cytotoxic effects at the skin level [15]. Mu et al. [16] reported size-dependent cellular uptake of GO in C2C12 cells, in which small nanosheets entered into cells mainly by clathrin-mediated endocytosis, and that larger sizes of graphene sheets enter into the cells through phagocytotic processes. Graphene-based nanomaterials (GBMs) induce apoptosis both in vivo and in vitro in a variety of cells and animals, based on surface area, layer number, lateral dimensions, functional groups, and surface chemistry [17]. For instance, GBMs induce apoptosis, cell cycle arrest, and DNA fragmentation, and also have effects on gonad development at a concentration of 10 mg/L in *Caenorhabditis elegans*. GBMs induce apoptosis through caspase activation, DNA fragmentation, increased oxidative stress, calcium efflux, loss of MMP, and decline in ATP synthesis [18–20]. Several studies reported that GBMs play significant roles in mitochondrial pathways, and mediate apoptosis via several signaling molecules such as Akt, MAPKs (JNK, ERK, and p38), and TGF-β. Gurunathan and co-workers reported the effects of GO and reduced GO in a variety of cancer and non-cancer cells. Their results demonstrated that reduced GO shows more toxicity than GO in cancer cells [21]. Conversely, nicotinamide-reduced GO displayed significant biocompatibility with mouse embryonic fibroblast cells [22].

Previously, several studies demonstrated that the effect of silver nanoparticles (AgNPs) induced toxicity via oxidative stress, apoptosis, and necrosis in mouse TM3 Leydig cells and mouse TM4 Sertoli cells [23,24]. AgNPs reduced the viability of spermatogonial stem cells (SSCs) in a dose and size-dependent manner, and also decreased the proliferation of SSCs by disrupting the glial cell-derived neurotrophic factor/Fyn kinase signaling pathways [23]. Furthermore, AgNPs reduced the expression of tight junction genes in TM4 cells, and of steroidogenesis-related genes in TM3 cells. Zhang et al. [24]

also demonstrated that AgNPs inhibit the growth of SSCs, with TM3 and TM4 cells treated as feeder cells for SSC culture.

Recently, several studies have been conducted to evaluate the cytotoxicity of different types of nanoparticles in various germ cells and also in animals, but none of these addressed the toxic effects of different sizes GO nanoparticles in Leydig cells (TM3) and Sertoli cells (TM4). Much less is known about the cytotoxic effects of GO on reproduction and development. In addition, a wide range of environmental factors have influences in different ways, including adverse health outcomes in later life that can be transmitted across multiple generations, giving rise to so-called multigenerational toxicity. Spermatogenesis is a complex developmental process supported by the secretion of hormones and other cellular signals from Leydig and Sertoli cells [23,24]. Toxic materials like graphene-induced damage to any of these main testicular cell types could reduce the production of healthy spermatozoa, impair fertility and/or adversely affect the resulting embryo.

To address this issue, we used a sonochemical method for the preparation of size-specified GO sheets, starting from large GO sheets, which is considered to be a top–bottom approach, and a relatively fast, low-cost, efficient and convenient method for preparing uniform GO nanosheets with sizes of 100 and 20 nm using normal graphite powder, which were used as starting material. Finally, we investigated the effect of two different sized sheets, with distinct mean size distributions of GO, on cytotoxicity, oxidative stress, DNA damage, and mechanisms of toxicity in Leydig cells (TM3) and Sertoli cells (TM4).

2. Results and Discussion

2.1. Synthesis and Characterization of Different Sizes of Graphene Oxide

GO was prepared by the Hummers' method [25], which involves the oxidation of graphite with strong oxidants and acids for a prolonged time. In this study, we prepared different sized GO particles by mild oxidation and a two-step centrifugation approach. The dispersion of different sized GO particles with the same concentration was characterized by UV–visible spectroscopy. The spectra for each sample were recorded, as shown in Figure 1A. Absorption peaks of GO-100 and GO-20 were observed at 233 and 227 nm. These peaks correspond to $\pi \rightarrow \pi^*$ transitions for C=C bonding, which is similar to that reported recently. A similar shoulder was also observed around 300 nm for GO-20, which is attributed to $n \rightarrow \pi^*$ transition of carbonyl groups (C=O). GO-20 showed a peak at about 227 nm (Figure 1B).

Next, we performed X-ray diffraction (XRD) to investigate the crystalline structures of GO sheets that were sonicated for different times. The characteristic peak of sonicated GO was measured and a strong peak appeared at $2\theta = 11.7°$ corresponding to an interlayer distance of 7.6 Å (d002). GO-100 and GO-20 displayed peaks at 11.7° and 11.88°, respectively (Figure 1C). All of these functional groups facilitate the hydration and exfoliation of graphene sheet in aqueous media. GO-20 shows a broad peak that can be fitted by using a Lorentzian function into single and sharp peaks centered at $2\theta = 11.88°$, corresponding to interlayer distances of 7.6 Å. These XRD results strongly suggest that graphite exfoliated significantly and converted into GO [26] (Figure 1D).

Further characterization was performed using Fourier-transform infrared spectroscopy (FTIR). FTIR was used to characterize the functional groups of the GO particles. The FTIR spectra of GO sonicated at different time points, with a variety of resultant sizes, are depicted in Figure 1E. Characteristic features of bands were observed for the sonicated GO-100 sample at 1639 cm^{-1} corresponding to C=O stretching vibrations of COOH groups, and at 3315 cm^{-1}, corresponding to C–O stretching vibrations of epoxy groups, and arising from -OH stretching vibrations. An increase in sonication intensity and time resulted in gradual positional changes of functional groups. FTIR spectroscopy further confirmed the GO-20. After further sonication, interesting peaks were observed in the spectrum of GO at 3320 and 1636 cm^{-1} corresponding to C=O stretching vibrations of COOH, groups which are attributed to C=O bonds in carboxylic acid and carbonyl moieties, respectively.

Interestingly, GO exhibits the characteristic features of the stretching vibration of C=O and OH groups, and other oxygen-containing functional groups in GO were also observed (Figure 1F).

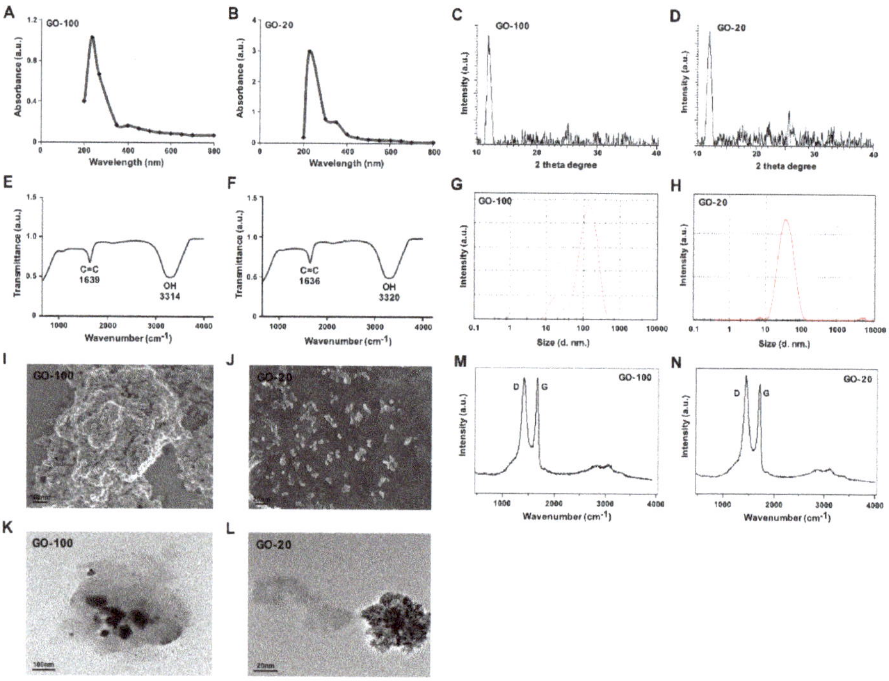

Figure 1. Synthesis and characterization of ultra-small GO. Ultraviolet-visible spectroscopy of GO-100 (**A**) and GO-20 (**B**). XRD images of GO-100 (**C**) and GO-20 (**D**). FTIR images of GO-100 (**E**) and GO-20 (**F**). Dynamic light-scattering (DLS) spectra of GO-100 (**G**) and GO-20 (**H**). SEM images of GO-100 (**I**) and GO-20 (**J**). TEM images of GO-100 (**K**) and GO-20 (**L**). Raman spectroscopy images of GO (**M**) and GO-20 (**N**). At least three independent experiments were performed for each sample and reproducible results were obtained. The data represent the results of a representative experiment.

Size distribution analysis is an important factor to determine either toxicity or biocompatibility analysis in aqueous solutions. Therefore, we determined the size of GO-100 and GO-20 using DLS with a concentration of 250 μg/mL. The average hydrodynamic diameter of GO-100 and GO-20 was 100 ± 10 nm and 20 ± 2 nm, respectively (Figure 1G,H). The hydrodynamic diameter and zeta potential of GO-100 and GO-20 in different dispersion media are shown in Table 1. Previous studies reported that the size of graphene and reduced graphene sheets as at least greater than 500 nm. For instance, GO reduced with isocyanate and carboxyl yielded average sizes of 560 ± 60 nm and 1110 nm respectively. Muthoosamy et al. [27] reported that by using mushroom extract and sonication, the average sizes of GO and rGO particles were 313 nm and 181 nm, respectively. That the smaller-sized rGO particles were obtained could be mainly due to the effect of 10 minutes of ultrasonication.

Next, we determined the size and surface morphology of GO-100 and GO-20 using SEM. As shown in Figure 1I,J, the SEM micrograph of GO shows layered flakes which are wrinkled, foamy, and creamy white. The presence of flakes implies that the graphene layers were fully oxidized to GO [28]. GO-20 exhibits crumpled, thin and smaller micro sheets which accumulated to form material with a disordered structure [8]. The size of the GO-100 and GO-20 were approximately 100 and 20 nm, respectively, which is in significant agreement with DLS and TEM.

Table 1. Hydrodynamic diameter and zeta potential of GO-100 and GO-20 in different dispersion media.

Name of the Sample	Hydrodynamic Diameter in Water	Hydrodynamic Diameter in DMEM Media	Hydrodynamic Diameter in DMEM Media +10% FBS
GO-100 (Size nm)	100	150	120
GO-100 (Zeta potential mV)	−50.8	−22.5	−13.5
GO-20 (Size nm)	20	40	30
GO-20 (Zeta potential mV)	−40.4	−12.8	−8.3

The size and morphology of GO-100 and GO-20 particles were determined using high-resolution TEM. TEM images of GO-100 and GO-20 display typical transparent morphology and rippled surfaces. The higher transparency areas indicate a few layers of reduced GO resulting from stacked nanostructure exfoliation [29]. The TEM images of GO-100 and GO-20 revealed small sizes of 100 and 20 nm, respectively (Figure 1K,L). Furthermore, GO particles showed significant levels of transparency, indicating a high degree of oxidation; the darker regions of GO represent the stacking of some GO layers [4,30].

Raman spectroscopy is a valuable tool to study the structural and electronic characterization of a variety of carbon materials. The Raman spectrum provides useful information on defects, carbon sp^2 vibrations, and the stacking order [31]. Figure 1M shows that the Raman spectrum of GO-100 shows a peak at 1585 cm^{-1}, related to the G band of the graphitic structure (sp^2 carbon), and D band located at 1345 cm^{-1}. As shown in Figure 1N, the Raman spectrum of GO-20 displayed significant peaks for D and G band positions. The Raman spectrum showed that the typical features of GO-20 are a D band at ~1360 cm^{-1}, and a G band at ~1587 cm^{-1}. The D band was assigned to the breathing mode of the k-point phonons with A1g symmetry, whereas the G band introduced the E2g phonon of the carbon sp^2 atoms [4,21]. It is also possible to observe a broad peak located at 2695 cm^{-1}, corresponding to the 2D band, which further confirmed the presence of a few layers of graphene sheets in GO-20. The presence of a large D mode indicates that the graphene flakes are rather defective.

2.2. GO-100 and GO-20 Inhibit Cell Viability in TM3 and TM4 Cells

The cytotoxicity of GO-100, GO-20, and the GO nanosheets was investigated. The synthesized GO-100 and GO-20 was dissolved in PBS at 1 mg/mL. Two different cell lines, TM3 and TM4, were incubated with the GO-100 and GO-20 dispersed in phosphate buffered saline (PBS) at various concentrations (0, 10, 20, 40, 60, 80, and 100 μg/mL) for a period of 24 h. In order to quantify the toxicity of GO-100 and GO-20, a CCK-8 assay was performed. The data from cell viability assays showed that both GO-100 and GO-20 exhibited dose-dependent effects on cell viability, and also revealed strong differences between GO-100- and GO-20-treated cells in terms of cell viability (Figure 2A,B). GO-20 treatment inhibited growth of TM3 and TM4 cells significantly more than GO-100, which was due to further exfoliation of GO by sonication, and also because of the smaller size of GO-20. Interestingly, TM3 cells were more sensitive than TM4, which is similar to previous reports which suggested that silver nanoparticles inhibit the viability of TM3 cells more than TM4 [24]. Akhavan et al. [12] reported size-dependent toxicity of reduced GO nanoplatelets (rGONPs) in mesenchymal stem cells (MSC). The data suggest size-dependent toxicity for both cell lines in a comparable manner, whereas reduction of GO also plays an important role. GO-100 and GO-20 nanosheets resulted in dose-dependent toxicity in TM3 and TM4 cells, with GO-20 being more cytotoxic than GO-100. Cho et al. [32] observed that cell viability was dependent on GO size and concentration, and cell viability was greater with smaller GOs such as single layer graphene oxide (SLGO) and medium layer graphene oxide (MLGO) than larger GOs. Moreover, SLGOs have higher cytotoxicity than MLGOs. Collectively, GO causes size and dose dependent effects in TM3 and TM4 cells. To corroborate the results obtained from cell viability and proliferation assays, next we assessed the cell morphology of TM3 and TM4 in the presence of GO-100 and GO-20 and also with AgNPs as a positive control. As a result of treatment, GO-100 and GO-20 treated cells exhibited flat to round morphology and an increased number of round cells (Figure 2C,D). The results suggest that the cells were undergoing morphological changes associated with apoptosis, such as cell shrinking and membrane. We observed a similar phenomenon in AgNPs treated cells.

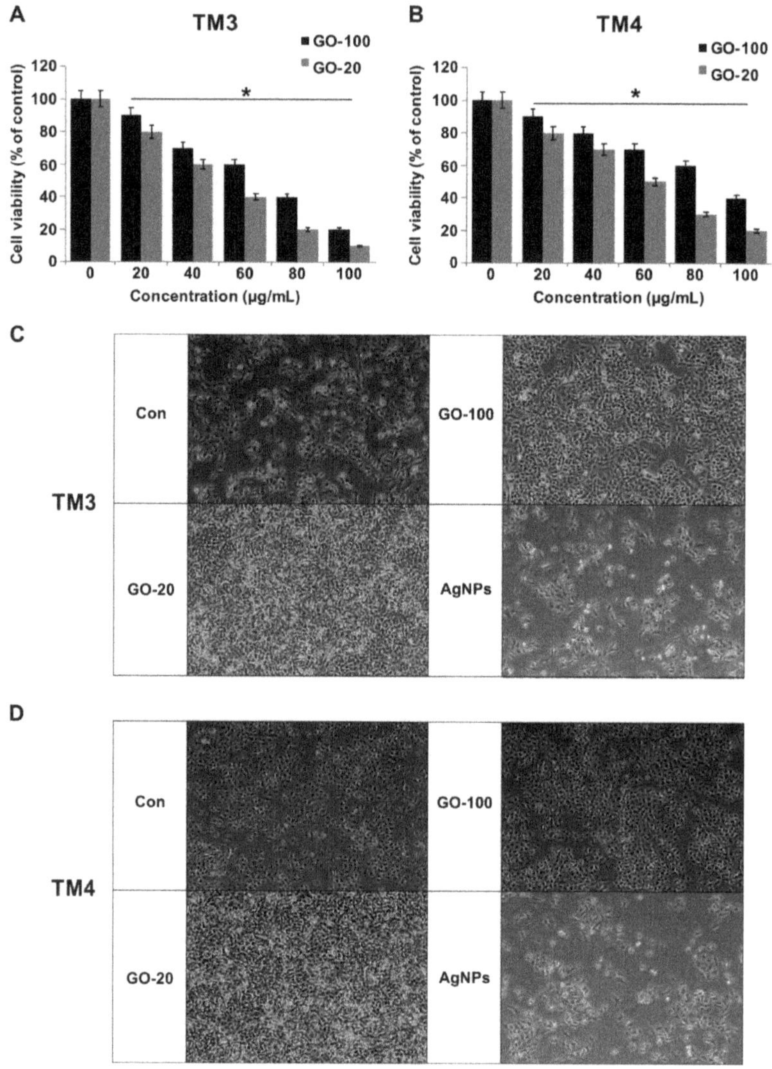

Figure 2. GO-100 and GO-20 graphene sheets inhibit viability of TM3 and TM4 cells. (**A**) The viability of TM3 cells was determined after 24 h exposure to different concentrations of GO-100 (20–100 µg/mL) and GO-20 (20–100 µg/mL), and (**B**) the viability of TM4 cells was determined after 24 h exposure to different concentrations of GO-100 (20–100 µg/mL) and GO-20 (20–100 µg/mL) using the CCK-8 assay. (**C**) Cell morphology of TM3 was assessed under a light microscope. (**D**) Cell morphology of TM4 was assessed under a light microscope. The results are expressed as the mean ± standard deviation of three independent experiments. At least three independent experiments were performed for each sample. The treated groups showed statistically significant differences from the control group by Student's *t*-test (* $p < 0.05$). Scale bar 200 µm

2.3. GO-100 and GO-20 Inhibit Proliferation of TM3 and TM4 Cells

Inhibition effects of GO-100 and GO-20 on cell proliferation in TM3 and TM4 cells were examined after GO-100 and GO-20 (0, 10, 20, 40, 60, 80, and 100 μg/mL) treatments (Figure 3A,B). GO-100 and GO-20 nanosheets resulted in dose-dependent toxicity in both TM3 and TM4 4 cells, with GO-20 being more cytotoxic than GO-100. The cell proliferation rate was profoundly decreased following treatment with 60 μg/mL GO-100 and GO-20, which resulted in 40% and 60% of the inhibitory effect observed in TM3 cells, respectively, whereas TM4 cells treated with 60 μg/mL of GO-100 and GO-20 resulted 30% and 50% of the inhibitory effect observed in TM4 cells. The degree of inhibition of the proliferation rate was more pronounced by GO-20 in both cell types, and TM3 cells exhibited more sensitivity than TM4 in both GO-100 and GO-20. Fiorillo et al. [33] demonstrated the proliferative effect of (small-GO) with flake sizes of 0.2–2 μm, and large GO (b-GO) with flake sizes of 5–20 μm, on six different type of cancer cells, including breast, ovarian, prostate, lung, pancreatic, and glioblastoma. The results drawn from this study suggest that GO effectively inhibits tumor formation. Among these two different types of GOs, small GO showed significant effects on the tested cell types, due to the ease of entry of small GO particles into the cells. Lioa et al. [34] found that smallest sized GO particles showed the greatest hemolytic activity, whereas aggregated graphene sheets exhibited the lowest hemolytic activity in human red blood cells. Choi et al. [35] reported that GO, rGO and GO silver nanocomposite significantly inhibit proliferation of subpopulations of OvCSCs, including ALDH+CD133+, ALDH+CD133−, ALDH−CD133 cells. GO-silver nanocomposite enhances differentiation of neuroblastoma cancer cells at low concentrations, and higher concentrations inhibit cell viability and proliferation [36]. Taken together, all these results suggest that GO inhibits cell proliferation, depending on the size and cell types involved.

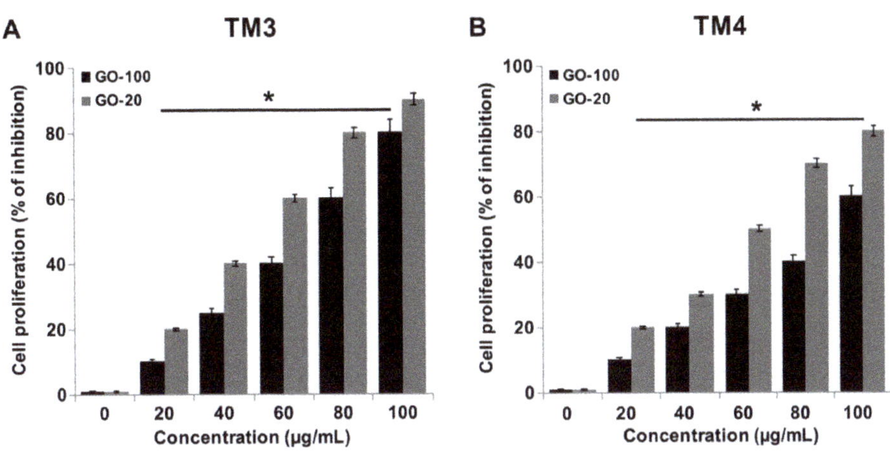

Figure 3. GO-100 and GO-20 graphene sheets inhibit proliferation of TM3 and TM4 cells. (**A**) The viability of TM3 cells was determined after 24 h exposure to different concentrations of GO-100 (20–100 μg/mL) and GO-20 (20–100 μg/mL), and (**B**) the viability TM4 cells was determined after 24 h exposure to different concentrations of GO-100 (20–100 μg/mL) and GO-20 (20–100 μg/mL) using the BrdU assay. The results are expressed as the mean ± standard deviation of three independent experiments. At least three independent experiments were performed for each sample. The treated groups showed statistically significant differences from the control group by Student's *t*-test (* $p < 0.05$).

2.4. Effect of GO-100 and GO-20 on LDH

Measuring lactate dehydrogenase activity is a good indicator for cell membrane damage and cytotoxicity. Graphene influences membrane integrity and dynamics via direct/indirect mechanisms in a variety of mammalian cells. Graphene can impair plasma membrane integrity and cause cell death. Therefore, we investigated the impact of GO-100 and GO-20 on LDH. TM3 and TM4 cells were treated with various concentrations of GO-100 and GO-20 for 24, and then the level of leakage of LDH was measured. The results indicated that GO-100 and GO-20 dose-dependently increase the leakage of LDH (Figure 4A,B). However, the leakage of LDH was significantly higher in GO-20 treated cells than GO-100. Interestingly, TM3 cells exhibited greater sensitivity to leakage of significant amounts of LDH in GO-100 and GO-20 treated TM3 cells, compared to TM4. Lammel et al. [37] observed that nano-sized graphene could influence the ultrastructure of the plasma membrane, resulting in a loss of membrane structural integrity. Pristine GO could impair cell membrane integrity and function by regulation of membrane- and cytoskeleton-associated genes, including *Actg2*, Myosin, *Tubb2a*, and Nebulin [38]. Li et al. reported that GLC-82 lung cancer cells treated with GO resulted in loss of plasma membrane integrity [39]. Similar results have been also observed in other cell types with GO, reduced GO, and GO-silver nanocomposite, which impairs the cell membrane integrity in a variety of cancer cells, including human breast cancer cells [40], human ovarian cancer cells [21] and human neuroblastoma cancer cells [30]. Internalization of graphene with endocytosis inhibitors attenuates the graphene-induced plasma membrane damage [41]. THP-1 cells treated with single-layered GO resulted in a dose-dependent higher level of leakage of LDH, compared to multilayered GO [42]. Taken together, GO-100 and GO-20 show significant effects on LDH containment, and eventually cause cytotoxicity.

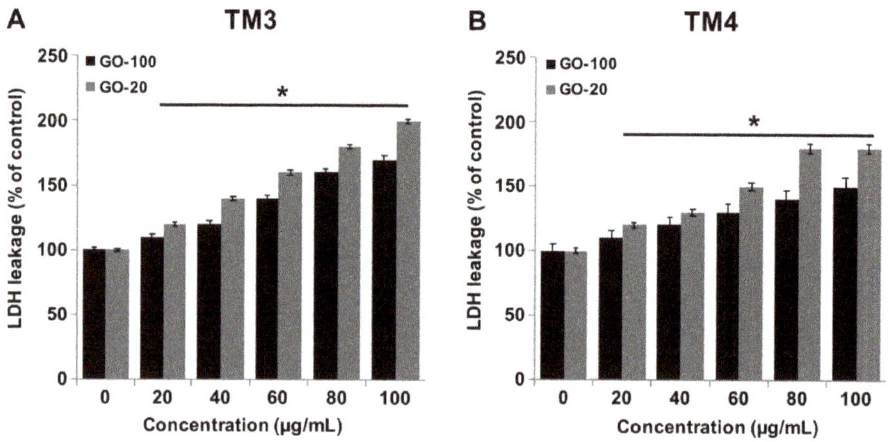

Figure 4. GO-100 and GO-20 graphene sheets increase the leakage of LDH. (**A**) TM3 cells were treated with GO (20–100 µg/mL) and GO-20 (20–100 µg/mL) for 24 h, and LDH activity was measured at 490 nm using an LDH cytotoxicity kit. (**B**) TM4 cells were treated with GO (20–100 µg/mL) and GO-20 (20–100 µg/mL) for 24 h, and LDH activity was measured at 490 nm using the LDH cytotoxicity kit. At least three independent experiments were performed for each sample. The results are expressed as the mean ± standard deviation of three independent experiments. The treated groups showed statistically significant differences from the control group by Student's *t*-test (* $p < 0.05$).

2.5. GO-100 and GO-20 Decrease MMP

To identify the effects of graphene materials on MMP, TM3 and TM4 cells were treated with various concentrations of GO-100 and GO-20 (20–100 µg/mL) for 24 h. Alterations to cell permeability after treatment with GO-100 and GO-20 were explored, as shown in Figure 5A,B. There was a significant difference between controls and cells exposed to GO-100 and GO-20, even at low-concentration of 20 µg/mL. Mitochondrial dysfunction is also associated with the overproduction of ROS [37]. Li et al. [20] reported that graphene causes a decrease in MMP and, consequently, increases levels of intracellular ROS, which activate the mitochondria-dependent apoptotic pathway. Nano-sized

GO, pristine graphene and graphene quantum dots also caused time- and concentration-dependent decreases in the MMP [41,43]. Our results showed significant agreement with previous reports suggesting that the small size of GO particles induces mitochondrial dysfunctions. Graphene alters calcium levels and decreases the MMP, and subsequently triggers apoptosis by the activation of mitochondria mediated mitogen-activated protein kinases (MAPKs) and transforming growth factor-beta (TGF-β)-related signaling pathways [20]. GO plays a significant role in ROS generation; for example, at low concentrations (<4 μg/mL), GO resulted in perturbation of mitochondrial structure and function in Hep G2 cells, whereas higher concentrations of graphene quantum dots (<200 μg/mL) also caused decreases in the MMP by increased ROS generation, in association with apoptotic and autophagic cell death, with an increase in the expression of caspase 3, caspase 9, beclin 1, and microtubule-associated protein 1A/1B-light chain 3 [43]. Taken together, GO causes elevated levels of loss of MMP, compared to GO-100, in TM3 and TM4 cells. In addition, the results suggested that TM3 cells show more sensitivity than TM4 cells.

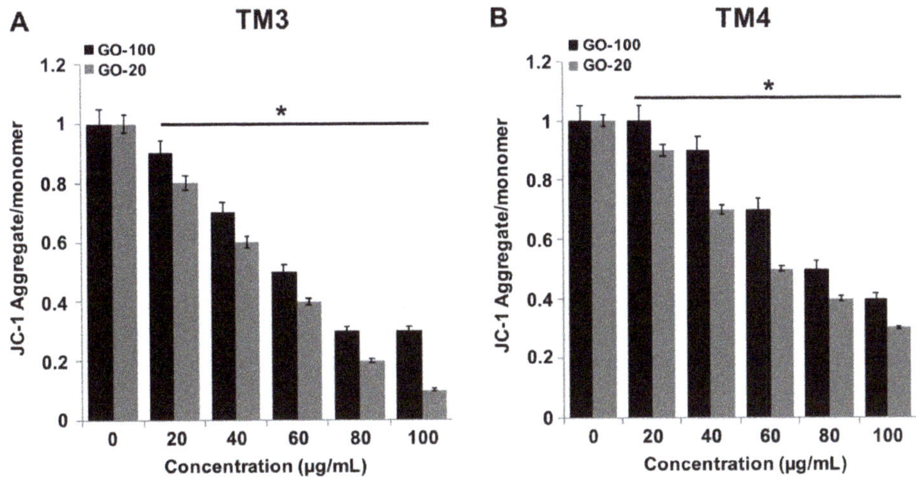

Figure 5. GO-100 and GO-20 graphene sheets impair mitochondrial membrane potential. (**A**) TM3 cells were treated with GO (20–100 μg/mL), (**B**) GO-20 (20–100 μg/mL) for 24 h, and the mitochondrial membrane potential (MMP) was determined using the cationic fluorescent indicator JC-1 (**B**). TM4 cells were treated with GO (20–100 μg/mL), (**B**) GO-20 (20–100 μg/mL) for 24 h, and MMP was determined using the cationic fluorescent indicator JC-1. The treated groups showed statistically significant differences from the control group by Student's *t*-test (* $p < 0.05$).

2.6. GO-100 and GO-20 Induce ROS Generation

GOs are able to induce ROS generation in a size-dependent manner. ROS generation is one of the primary mechanisms of nanoparticle-induced toxicity and apoptosis [4]. ROS play a key role in both the extrinsic and intrinsic pathways of apoptosis. Therefore, it is necessary to determine the effects of GO-100 and GO-20 on ROS generation. Exposure of TM3 and TM4 cells to various concentrations GO-100 and GO-20 induced size-dependent oxidative stress, as indicated by the degree of ROS generation (Figure 6A,B). ROS generation increased more rapidly with GO-20 than with GO-20. Pelin et al. [44] reported that few-layer-graphene (FLG) and GO induce time- and concentration-dependent cellular ROS production in human HaCaT skin keratinocytes. GO significantly increased ROS production at concentrations of 33 μg/mL and above in the tested assays. Similarly, our findings also accorded with those of Pelin et al. [44]; we found that above 30 μg/mL and above yielded significant levels of ROS generation. Gurunathan and coworkers reported that GO and rGO dose-dependently produced significant level of ROS in a variety of human cell lines, including breast cancer cells [40], ovarian cancer cells [21] and neuroblastoma cancer cells [30]. ROS can cause some adverse effects, including downregulation of defensive systems to disrupt the structure and function of normal cells, and to cause unbalanced levels of antioxidants. ROS can also cause damage to macromolecules such as lipids, proteins and DNA, further resulting in increasing levels of release of inflammatory cytokines and chemokines [45]. Intracellular accumulation of ROS increases apoptosis in murine RAW 264.7 macrophages at concentrations of 20–100 μg/mL. Chang et al. [13] reported that GO causes concentration-dependent toxicity in A549 cells at levels of 200 μg/mL, and eventually leads to reduction in cell viability, whereas low concentrations of GO (10 μg/mL) did not enter A549 cells, and had no obvious toxicity.

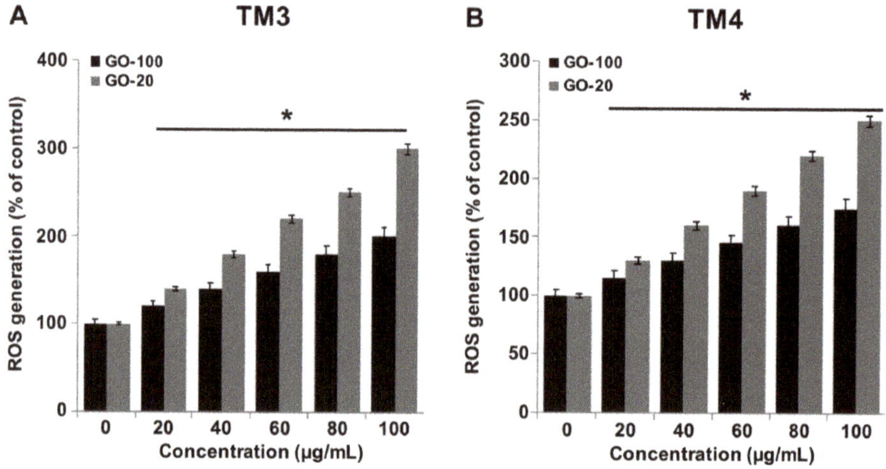

Figure 6. GO-100 and GO-20 graphene sheets induce ROS generation. (**A**) TM3 cells were treated with GO (20–100 μg/mL) and GO-20 (20–100 μg/mL) for 24 h and ROS levels were measured using DCFH-DA. (**B**) TM3 cells were treated with GO (20–100 μg/mL) and GO-20 (20–100 μg/mL) for 24 h and ROS levels were measured using DCFH-DA. Relative fluorescence of 2′,7′-dichlorofluorescein was measured at an excitation wavelength of 485 nm and emission wavelength of 530 nm using a spectrofluorometer. The treated groups showed statistically significant differences from the control group by Student's *t*-test (* $p < 0.05$).

2.7. GO-100 and GO-20 Cause DNA Damage in TM3 and TM4 Cells

GO nanosheets have potential to induce *in vivo* and *in vitro* mutagenesis [46,47]. To evaluate the effects of GO-100- and GO-20-induced ROS on DNA damage in TM3 and TM4 cells, 8-oxo-dG levels were determined. A 24 h exposure of TM3 and TM4 cells to various concentrations of GO-100 and GO-20 increased oxidative DNA damage, as indicated by a significant elevation of 8-oxo-dG production. The results indicated that GO-100 and GO-20 dose-dependently increased the level of 8-OhdG (Figure 7A,B). A significant difference was observed in DNA damage by measuring the levels of 8-oxo-dG after exposure to GO-100 and GO-20 for 24 h, compared to the control groups. Interestingly, the level of 8-OhdG was significantly higher in GO-20 treated cells than GO-100. Interestingly, TM3 cells exhibited greater sensitivity, both to GO-100 and GO-20, compared to TM4, with increased levels of 8-OhdG. Chatterjee et al. found graphene-family nanomaterials (GFNs) caused DNA damage to human bronchial epithelial cells [41]. In agreement with these findings, our results showed that both sizes of GO-100 and GO-20 could induce cytotoxic effects in TM3 and TM4 cells, and exposure to higher concentrations of GO (50–100 μg/mL) significantly reduced cell viability, increased cytotoxicity, and induced DNA damage in TM3 and TM4 cells, which consequently altered the gene expression levels of various pro- and anti-apoptotic genes. Studies from our group and other research groups confirmed that GO exposure generated ROS at cellular levels, which could be the principal mechanism of GO in inducing DNA damage and genotoxicity [3,28,48,49]. Previous studies reported that GO exposure affects the activity of LDH, induces cell immune toxicity, and generates ROS, which ultimately lead to oxidative stress and DNA damage [50,51]. Collectively, our findings demonstrated that GO-100 and GO-20 caused DNA damage in TM3 and TM4 cells, which ultimately caused cell death.

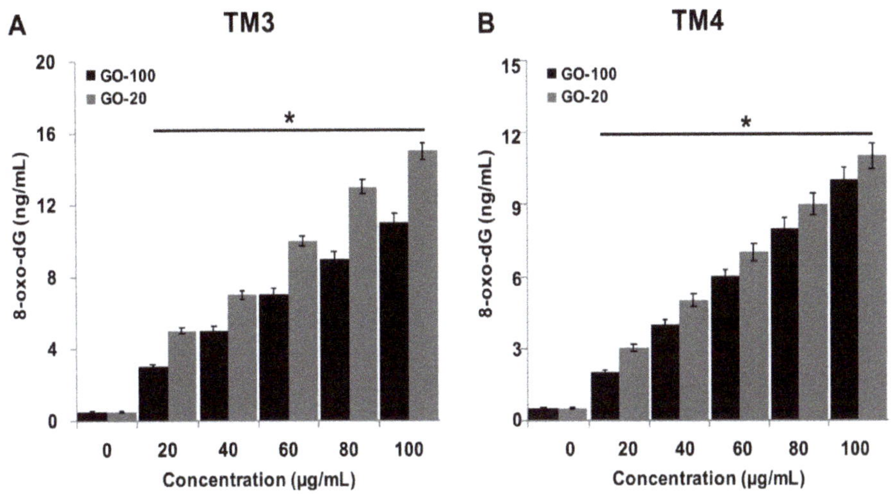

Figure 7. GO-100 and GO-20 graphene sheets induce DNA damage. (**A**) TM3 were treated cells with GO (20–100 μg/mL) and GO-20 (20–100 μg/mL) for 24 h and ROS levels were measured using DCFH-DA. (**B**) TM3 cells were treated with GO (20–100 μg/mL) and GO-20 (20–100 μg/mL) for 24 h and 8-oxo-dG levels were measured. The treated groups showed statistically significant differences from the control group by Student's *t*-test (* $p < 0.05$).

2.8. ROS Increase Upregulation of Pro-Apoptotic Genes and Downregulation of Anti-Apoptotic Genes in TM3 and TM4 Cells

Several studies have reported that the mechanisms of GO- and graphene oxide silver nanocomposite-induced toxicity in numerous human cancer cell lines include oxidative stress, DNA damage, and apoptosis [21]. To further substantiate size-dependent toxicity of GO-100- and GO-20-induced DNA damage in TM3 and TM4 cells, we measured the expression of representative genes responsible for DNA damage repair and apoptosis in GO-100 and GO-20 treated and untreated cells. ROS maintains equilibrium in cells with a variety of antioxidant molecules. ROS are molecules considered to be a double-sided coin. A lower level of ROS is essential for normal physiological functions [52]. To determine the effect of GO-100 and GO-20, and the role of ROS in terms of apoptotic cell fate, TM3 and TM4 cells were treated with GO-100 and GO-20 (50 μg/mL) for 24 h. We measured relative expression of *p53, p21, Bax, Bak, caspase-3* and *Bcl-2*. The results showed that all the tested genes were upregulated except *Bcl-2*. All of the upregulated genes, such as *p53, p21, Bax, Bak,* and *caspase-3*, could increase the levels of ROS, which is responsible for the upregulation of pro-apoptotic and down regulation of anti-apoptotic genes in TM3 and TM4 cells (Figure 8).

Similarly, Yuan and Gurunathan [53] reported that GO nanocomposite upregulated *P53, P21, BAX, BAK*, and *CASP3* in human cervical cancer cells. Interestingly, GO-20 increased the upregulation of all tested genes in both TM3 and TM4 cells; however, the expression levels were higher in TM3 cells compared to TM4. Furthermore, TM3 seems to be more sensitive than TM4, as related to expression of pro-apoptotic genes. Zhang et al [24] demonstrated that AgNPs regulate two different signaling pathways in TM3 and TM4 cells. TM3 cells treated with 55 μg/mL of 10 nm and 20 nm AgNPs upregulated the expression p53, p-Erk1/2, Bax, Bcl-2, and RAD51, whereas in TM4 cells, p53 and p38 phosphorylation was not significantly affected, whereas the phosphorylation of Erk1/2, Bax, Bcl2, and RAD51 increased. The p53 tumor suppressor plays critical roles in cell cycle arrest and apoptosis through Bax, and Bak transactivation, mitochondrial cytochrome c release, and by caspase-9 activation, which is usually followed by the activation of caspase-3, -6, and -7 [54,55]. P21 is a direct downstream gene of p53. The activation of p53 induces expression of p21, which in turn induces cell cycle arrest and DNA damage. Our results showed that TM3 and TM4 cells treated with GO-100 and GO-20 showed 2-fold higher expression of p53. The increased expression levels of Bax, Bak, and caspase-3, and decreased expression levels of Bcl-2 by GO-100 and GO-20 further substantiate mitochondrial-mediated apoptosis in TM3 and TM4 cells [24]. Collectively, the toxic effects of GO-100 and GO-20 on testicular cell lines was mediated by inducing ROS production, and the consequent modulation of pro- and anti-apoptotic gene expression levels, reduction of MMP, and induction of DNA damage, leads to apoptosis.

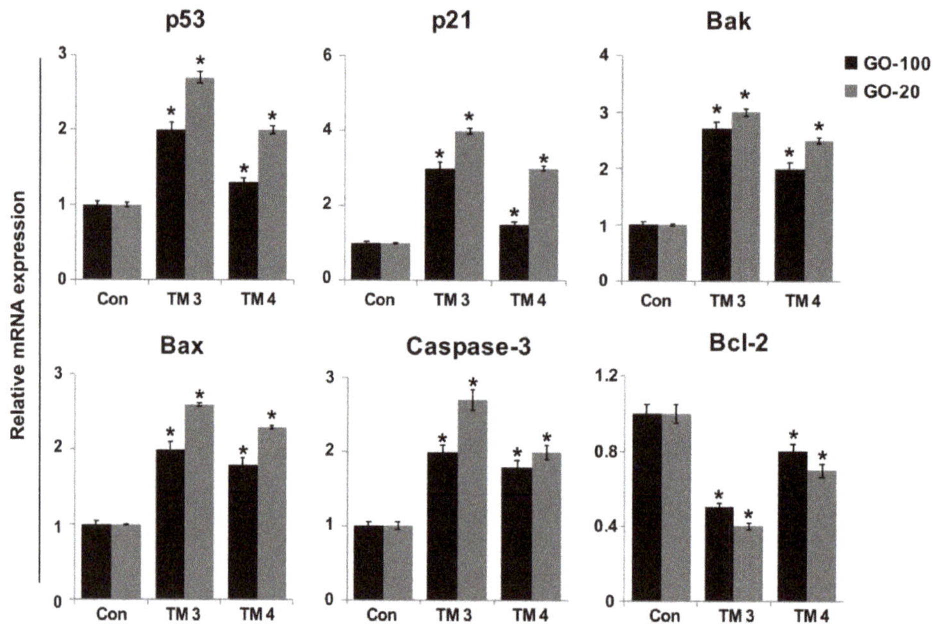

Figure 8. Effect of GO-100 and GO-20 graphene sheets on expression of pro- and anti-apoptotic genes. TM3 and TM4 cells were treated with GO (20–100 μg/mL) and GO-20 (20–100 μg/mL) for 24 h. After 24 h treatment, expression fold- levels were determined as fold- changes in reference to expression values of *GAPDH*. Results are expressed as fold changes. At least three independent experiments were performed for each sample. The treated groups showed statistically significant differences from the control group by Student's *t*-test (* $p < 0.05$).

2.9. GO-100 and GO-20 Nanosheets Reduced Phosphorylation Level of EGFR and AKT

The epidermal growth factor receptor (EGFR) is involved in pathogenesis, therapy, and prognosis, and is also involved in cell differentiation, proliferation, apoptosis, migration, and adhesion of various tumor types, including germ cells. Testicular germ cell tumors (TGCTs) are the most common type of cancer in young adult males [56,57]. Targeting EGFR signal transduction is necessary for the treatment of patients with TGCTs. Therefore, we analyzed the EGFR/AKT signaling pathway by assessment of phosphorylation levels of EGFR and AKT protein. Treatment with GO-100 and GO-20 decreased levels of phospho-EGFR, whereas the same did not alter total levels of β-actin protein (Figure 9A). The nanoparticles also decreased the activation of two downstream targets of EGFR and

AKT. The results clearly indicated that both GO-100 and GO-20 downregulate phosphorylation of AKT, both in TM3 and TM4 cells, compared to a control group. Our results showed altered phosphorylation levels of EGFR signaling molecules after GO-100 and GO-20 exposure. In GO-20 treated cells, the phosphorylation levels of EGFR showed a significant decrease compared to TM4, which is in agreement with all the cellular assays.

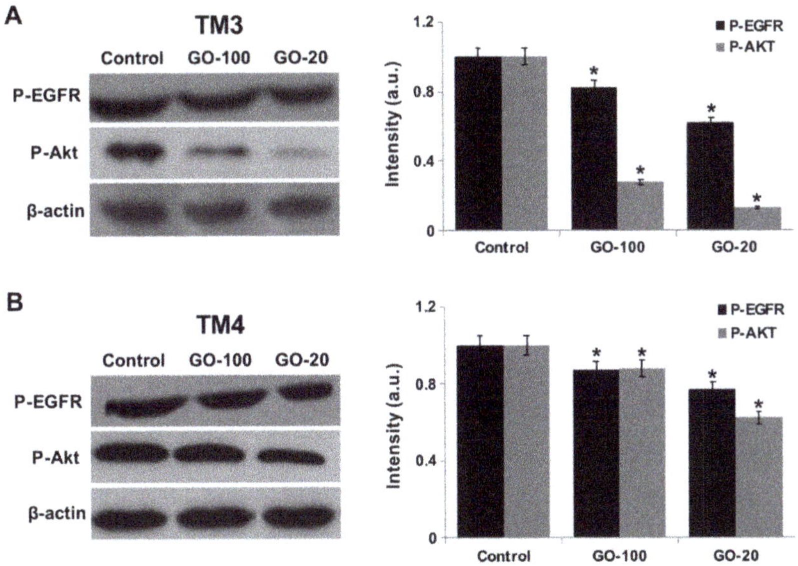

Figure 9. GO-100 and GO-20 graphene sheets alter phosphorylation levels of EGFR/AKT. (**A**) TM3 cells were treated with GO-100 and GO-20 at a dose of 50 μg/mL for 30 min. Total proteins were extracted, and the phosphorylation levels of EGFR/AKT signaling pathway molecules were analyzed via western blot. (**B**) TM4 cells were treated with GO-100 and GO-20 at a dose of 50 μg/mL for 30 min. Total proteins were extracted, and the phosphorylation levels of EGFR/AKT signaling pathway molecules were analyzed via western blot. The results are presented as the mean ± SEM from three independent experiments. The treated groups showed statistically significant differences from the control group by Student's *t*-test (* $p < 0.05$).

Several studies have indicated that nanomaterials trigger generation of ROS, which eventually alters specific biochemical and biological responses required for apoptosis through downregulation of phosphorylation of AKT. Akt/PKB is a serine/threonine protein kinase that functions as a critical regulator of cell survival and proliferation [58]. To determine the effect of GO-100 and GO-20 on phosphorylation of AKT, cells were incubated with 50 μg/ml of GO-100 and GO-20 for 30 min, and then proteins were extracted and probed against Phospho AKT and phosphorylation levels were determined. The results clearly indicated that both GO-100 and GO-20 downregulate phosphorylation of AKT, both in TM3 and TM4 cells, compared to controls (Figure 9B). Our results showed altered phosphorylation levels of AKT signaling molecules after GO-100 and GO-20 exposure. In GO-20 treated cells, the phosphorylation levels of AKT showed a significant decrease, followed by a decrease to baseline levels (p-AKT), or to levels lower than that of the total AKT. We attribute these interesting results to the exposure to a toxic dose (50 μg/mL) of GO-100 and GO-20. Our results indicated that GO-100 and GO-20 could induce significant apoptosis via the inactivation of AKT. Moreover, these GO-100- and GO-20-induced effects on inactivation are likely involved in changes throughout the cell death pathway. Interestingly, in GO-20 treated cells, the phosphorylation levels of AKT showed a significant decrease compared to TM4, which is in agreement with all the cellular assays. Furthermore,

P-AKT was significantly reduced in TM3 compared to TM4 cells. Collectively, the suppression of phosphorylation of EGFR and AKT leads to negative regulation of cell survival.

Graphene is able to cause cellular toxicity through ROS mediated signaling pathways. Graphene could affects cellular and biological behavior at the cellular, subcellular, protein and gene levels [45–49]. The possible potential stronger effect of toxicity of GO-20 compared to GO-100 is due to multiple factors including size, shape, concentration, lateral dimension, surface structure, surface chemistry, functional groups, purity, protein corona, adsorption and interaction with cells. Particularly, the important factor could be higher level of uptake rate of GO-20 compared to GO-100 which could eventually cause significant viability loss and oxidative stress. Generally, Graphene-based nanomaterials can enter into the cells via different pathways such as clathrin/caveolar-mediated endocytosis, phagocytosis, macropinocytosis, and pinocytosis and exeunt the cell via the pathways of lysosome secretion, vesicle-related secretion, and non-vesicle-related secretion. This interaction may lead to the possibility of events such as adsorption or incorporation of graphene onto the surfaces of cells. Furthermore, the entrapped biomolecules on the surface of graphene could influence the tertiary structure of a protein, resulting in the formation of a protein-graphene interface and malfunction. Graphene-induced ROS may cause oxidative stress, loss of cell function, mitochondrial damage, initiation of lipid peroxidation, covalent chemical modifications of nucleic acids, DNA-strand breaks, induction of gene expression via the activation of transcription factors, and modulation of inflammation via signal transduction, leading to toxicity, cell death and genotoxicity (Figure 10).

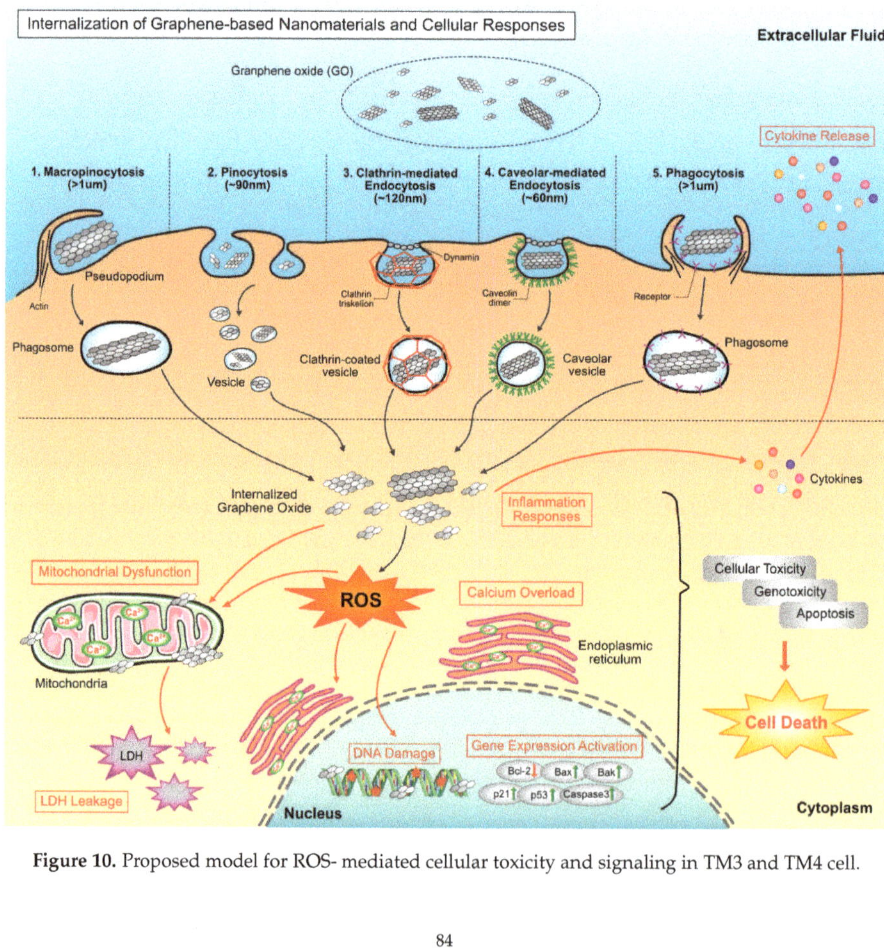

Figure 10. Proposed model for ROS- mediated cellular toxicity and signaling in TM3 and TM4 cell.

3. Materials and Methods

3.1. Preparation and Characterization of Different Sizes of GO Nanosheets

Graphene sheets were synthesized by a modification of Hummer's method [10,30,59,60]. Characterization of GO nanosheets was performed according to a method described previously [54].

3.2. Cell Culture, Cell Viability, Cell Proliferation and Measurement of LDH and ROS

TM3 (KCLB No 21714) and TM4 (KCLB No 21715) cell lines were obtained from Korean cell line bank (Seoul, South Korea). TM3 and TM4 cells were cultured in Dulbecco's Modified Eagle's Medium (DMEM) (Hyclone, Logan, UT, USA) supplemented with 10% fetal bovine serum (FBS), 100 U/mL penicillin, and 100 μg/mL streptomycin at 37 °C in a 5% CO_2 atmosphere. Cells were seeded onto 6-well plates at a density of 0.3×10^6 cells per well and incubated for 24 h prior to the experiments. The cells were washed with phosphate-buffered saline (PBS; pH 7.4) and incubated in fresh medium containing different concentrations of GO prepared in water.The CCK-8, cell proliferation, LDH and ROS assays were performed as described previously [35].

3.3. Measurement of 8-Oxo-dG

8-oxo-dG was determined as described previously [61], and also according to the kit manufacturer's instructions (Trevigen, Gaithersburg, MD, USA).

3.4. Mitochondrial Membrane Potential

MMP was measured as per the manufacturer's instructions (Molecular Probes, Eugene, OR, USA), and as described previously [62] using a cationic fluorescent indicator JC-1 (Molecular Probes).

3.5. Reverse Transcription-Quantitative Polymerase Chain Reaction (RT-qPCR) Assay

Total RNA was extracted from cells treated with GO-100 and GO-20 for 24 h using the Arcturus PicoPure RNA isolation kit (Arcturus Bioscience, Mountain View, CA, USA); the samples were then prepared according to the manufacturer's instructions. The lists of primers were shown in Table 2.

Table 2. List of primers used for quantitative real-time polymerase chain reaction for analysis of apoptotic gene expression.

Gene	Primer	TM (°C)
P53	F:AGAGACCGTACAGAAGA	58
	R:CTGTAGCATGGGATCCTTT	
P21	F:GTTGCTGTCCGGACTACCG	53
	R:AAAAACAATGCCACCACTCC	
Caspase-3	F:AGGGGTCATTTATGGGACA	58
	R:TACACGGGATCTGTTTCTTTG	
	R:CAGGCCTGGATGAAGAAGAG	
Bax	F:CGAGCTGATCAGAACCATCA	58
	R:GAAAAATGCCTTTCCCCTTC	
	R:AACCATACTCGAACCACATCCT	
Bcl-2	F:TAAGCTGTCACAGAGGGGCT	58
	R:TGAAGAGTTCCTCCACCACC	
	R:AAAGGAGGCTACACCCCAGT	
Bak	F: CTC AGA GTT CCA GAC CAT GTT G	58
	R: CAT GCT GGT AGA CGT GTA GGG	
	R:CCTTTGTACCGTTGCATCCT	
GAPDH	F:AGGTCGGTGTGAACGGATTTG	58

3.6. Western Blotting

Western blot was carried out according to method described earlier [24].

3.7. Statistical Analyses

All assays were conducted in triplicate, and each experiment was repeated at least three times. The results are presented as means ± standard deviation. All of the experimental data were compared using Student's *t*-test. A p-value less than 0.05 was considered statistically significant.

4. Conclusions

Graphene has promising effects in various biomedical applications such as antibacterial treatment, cancer diagnosis and therapy, gene delivery, and bio-imaging, due to its unique physical, chemical, biological and mechanical properties. Studies have explored its size-dependent cytotoxicity in a variety of cancer cells; however, there are no studies reported regarding the biological effects of ultra-small graphene nanosheets in germ cells such as Leydig (TM3) and Sertoli cells (TM4). To investigate for the first time size-dependent cyto and genotoxic effects of GO in germ cells, we synthesized uniform different sizes graphene oxide nanosheets with an average size of 100 and 20 nm, using a modified Hummers' method, and characterized them by various analytical techniques. Our results indicated that both GO-100 and GO-20 exert potent cytotoxic effects in a size- and dose- dependent manner, as evidenced by reduced cell viability, proliferation, leakage of LDH, generation of ROS, loss of MMP, and DNA damage. Particularly, both GO-100 and GO-20 showed significant effect in all cellular assays tested at high concentration (20–100 μg/mL) whereas there is no significant effect was observed at low concentration (1–20 μg/mL). Interestingly, GO-20 is able to cause significant intracellular ROS production, which in turn induces significant loss of MMP, upregulation of pro-apoptotic genes, and downregulation of anti-apoptotic genes, compared to GO-100. Further, exposure of TM3 and TM4 cells to GO-100 and GO-20 revealed increased levels of 8-oxo-dG, a typical oxidative marker for DNA damage. Furthermore, the degree of cytotoxicity differed between these two different GOs, possibly due to differences in size and shape, solubility and dispersion, entry, protein adsorption, particle surface characteristics, surface charge, surface-associated chemical groups, ions released from graphene, aggregation, and mode of interaction with cells. Moreover, use of GOs in biomedical applications should be investigated thoroughly to enhance its biocompatibility and also commercially available carbon based products need to be evaluated. We hope that the results of this study may represent a significant step forward to identify harmful effects of GOs in germ cells, and to develop safer nanomaterials. Future studies are required to enhance the biocompatibility of graphene-based materials for biomedical applications.

Author Contributions: S.G. originated the study concept, and participated in the design, experiments performed, analyzing data and writing of the manuscript. M.-H.K. and M.J. performed cellular assays. J.-H.K. provided facility and monitored all of the work done. All authors have read and approved the final manuscript.

Funding: This research received no external funding.

Acknowledgments: This paper was supported by Konkuk University in 2018.

Conflicts of Interest: The authors declare no conflict of interest.

References

1. Novoselov, K.S.; Geim, A.K.; Morozov, S.V.; Jiang, D.; Zhang, Y.; Dubonos, S.V.; Grigorieva, I.V.; Firsov, A.A. Electric field effect in atomically thin carbon films. *Science* **2004**, *306*, 666–669. [CrossRef]
2. Geim, A.K. Graphene: Status and prospects. *Science* **2009**, *324*, 1530–1534. [CrossRef]
3. Sanchez, M.D.; Chen, P.R.; Reinecke, T.; Muhler, M.; Xia, W. The Role of Oxygen- and Nitrogen-containing Surface Groups on the Sintering of Iron Nanoparticles on Carbon Nanotubes in Different Atmospheres. *ChemCatChem* **2012**, *4*, 1997–2004. [CrossRef]

4. Gurunathan, S.; Kim, J.H. Synthesis, toxicity, biocompatibility, and biomedical applications of graphene and graphene-related materials. *Int. J. Nanomed.* **2016**, *11*, 1927–1945. [CrossRef]
5. Avouris, P.; Dimitrakopoulos, C. Graphene: Synthesis and applications. *Mater. Today* **2012**, *15*, 86–97. [CrossRef]
6. Subrahmanyam, K.S.; Panchakarla, L.S.; Govindaraj, A.; Rao, C.N.R. Simple Method of Preparing Graphene Flakes by an Arc-Discharge Method. *J. Phys. Chem. C* **2009**, *113*, 4257–4259. [CrossRef]
7. Kim, K.S.; Zhao, Y.; Jang, H.; Lee, S.Y.; Kim, J.M.; Kim, K.S.; Ahn, J.H.; Kim, P.; Choi, J.Y.; Hong, B.H. Large-scale pattern growth of graphene films for stretchable transparent electrodes. *Nature* **2009**, *457*, 706–710. [CrossRef]
8. Stankovich, S.; Dikin, D.A.; Piner, R.D.; Kohlhaas, K.A.; Kleinhammes, A.; Jia, Y.; Wu, Y.; Nguyen, S.T.; Ruoff, R.S. Synthesis of graphene-based nanosheets via chemical reduction of exfoliated graphite oxide. *Carbon* **2007**, *45*, 1558–1565. [CrossRef]
9. Kosynkin, D.V.; Higginbotham, A.L.; Sinitskii, A.; Lomeda, J.R.; Dimiev, A.; Price, B.K.; Tour, J.M. Longitudinal unzipping of carbon nanotubes to form graphene nanoribbons. *Nature* **2009**, *458*, 872–876. [CrossRef]
10. Zhang, H.; Peng, C.; Yang, J.; Lv, M.; Liu, R.; He, D.; Fan, C.; Huang, Q. Uniform different sizes graphene oxide nanosheets with low cytotoxicity and high cellular uptake. *ACS Appl. Mater. Interfaces* **2013**, *5*, 1761–1767. [CrossRef]
11. Zhang, Y.; Ali, S.F.; Dervishi, E.; Xu, Y.; Li, Z.; Casciano, D.; Biris, A.S. Cytotoxicity effects of graphene and single-wall carbon nanotubes in neural phaeochromocytoma-derived PC12 cells. *ACS Nano* **2010**, *4*, 3181–3186. [CrossRef]
12. Akhavan, O.; Ghaderi, E.; Akhavan, A. Size-dependent genotoxicity of graphene nanoplatelets in human stem cells. *Biomaterials* **2012**, *33*, 8017–8025. [CrossRef]
13. Chang, Y.; Yang, S.T.; Liu, J.H.; Dong, E.; Wang, Y.; Cao, A.; Liu, Y.; Wang, H. In vitro toxicity evaluation of graphene oxide on A549 cells. *Toxicol. Lett.* **2011**, *200*, 201–210. [CrossRef]
14. Mullick Chowdhury, S.; Dasgupta, S.; McElroy, A.E.; Sitharaman, B. Structural disruption increases toxicity of graphene nanoribbons. *J. Appl. Toxicol.* **2014**, *34*, 1235–1246. [CrossRef]
15. Pelin, M.; Fusco, L.; Leon, V.; Martin, C.; Criado, A.; Sosa, S.; Vazquez, E.; Tubaro, A.; Prato, M. Differential cytotoxic effects of graphene and graphene oxide on skin keratinocytes. *Sci. Rep.* **2017**, *7*, 40572. [CrossRef]
16. Mu, Q.; Su, G.; Li, L.; Gilbertson, B.O.; Yu, L.H.; Zhang, Q.; Sun, Y.P.; Yan, B. Size-dependent cell uptake of protein-coated graphene oxide nanosheets. *ACS Appl. Mater. Interfaces* **2012**, *4*, 2259–2266. [CrossRef]
17. Ou, L.; Lin, S.; Song, B.; Liu, J.; Lai, R.; Shao, L. The mechanisms of graphene-based materials-induced programmed cell death: A review of apoptosis, autophagy, and programmed necrosis. *Int. J. Nanomed.* **2017**, *12*, 6633–6646. [CrossRef]
18. Sasidharan, A.; Swaroop, S.; Chandran, P.; Nair, S.; Koyakutty, M. Cellular and molecular mechanistic insight into the DNA-damaging potential of few-layer graphene in human primary endothelial cells. *Nanomed. Nanotechnol. Biol. Med.* **2016**, *12*, 1347–1355. [CrossRef]
19. Choi, Y.J.; Kim, E.; Han, J.W.; Kim, J.H.; Gurunathan, S. A Novel Biomolecule-Mediated Reduction of Graphene Oxide: A Multifunctional Anti-Cancer Agent. *Molecules* **2016**, *21*, 375. [CrossRef] [PubMed]
20. Li, Y.; Liu, Y.; Fu, Y.J.; Wei, T.T.; Le Guyader, L.; Gao, G.; Liu, R.S.; Chang, Y.Z.; Chen, C.Y. The triggering of apoptosis in macrophages by pristine graphene through the MAPK and TGF-beta signaling pathways. *Biomaterials* **2012**, *33*, 402–411. [CrossRef]
21. Gurunathan, S.; Han, J.W.; Park, J.H.; Kim, E.; Choi, Y.J.; Kwon, D.N.; Kim, J.H. Reduced graphene oxide-silver nanoparticle nanocomposite: A potential anticancer nanotherapy. *Int. J. Nanomed.* **2015**, *10*, 6257–6276. [CrossRef]
22. Zhang, X.F.; Liu, Z.G.; Shen, W.; Gurunathan, S. Silver Nanoparticles: Synthesis, Characterization, Properties, Applications, and Therapeutic Approaches. *Int. J. Mol. Sci.* **2016**, *17*, 1534. [CrossRef]
23. Braydich-Stolle, L.K.; Lucas, B.; Schrand, A.; Murdock, R.C.; Lee, T.; Schlager, J.J.; Hussain, S.M.; Hofmann, M.C. Silver nanoparticles disrupt GDNF/Fyn kinase signaling in spermatogonial stem cells. *Toxicol. Sci.* **2010**, *116*, 577–589. [CrossRef]
24. Zhang, X.F.; Choi, Y.J.; Han, J.W.; Kim, E.; Park, J.H.; Gurunathan, S.; Kim, J.H. Differential nanoreprotoxicity of silver nanoparticles in male somatic cells and spermatogonial stem cells. *Int. J. Nanomed.* **2015**, *10*, 1335–1357.

25. Hummers, W.S., Jr.; Offeman, R.E. Preparation of graphitic oxide. *J. Am. Chem. Soc.* **1958**, *80*, 1339. [CrossRef]
26. McAllister, M.J.; Li, J.-L.; Adamson, D.H.; Schniepp, H.C.; Abdala, A.A.; Liu, J.; Herrera-Alonso, M.; Milius, D.L.; Car, R.; Prud'homme, R.K. Single sheet functionalized graphene by oxidation and thermal expansion of graphite. *Chem. Mater.* **2007**, *19*, 4396–4404. [CrossRef]
27. Muthoosamy, K.; Bai, R.G.; Abubakar, I.B.; Sudheer, S.M.; Lim, H.N.; Loh, H.S.; Huang, N.M.; Chia, C.H.; Manickam, S. Exceedingly biocompatible and thin-layered reduced graphene oxide nanosheets using an eco-friendly mushroom extract strategy. *Int. J. Nanomed.* **2015**, *10*, 1505–1519.
28. Gurunathan, S.; Han, J.W.; Dayem, A.A.; Eppakayala, V.; Kim, J.H. Oxidative stress-mediated antibacterial activity of graphene oxide and reduced graphene oxide in Pseudomonas aeruginosa. *Int. J. Nanomed.* **2012**, *7*, 5901–5914. [CrossRef]
29. Stobinski, L.; Lesiak, B.; Malolepszy, A.; Mazurkiewicz, M.; Mierzwa, B.; Zemek, J.; Jiricek, P.; Bieloshapka, I. Graphene oxide and reduced graphene oxide studied by the XRD, TEM and electron spectroscopy methods. *J. Electron Spectrosc. Relat. Phenom.* **2014**, *195*, 145–154. [CrossRef]
30. Gurunathan, S.; Kim, J.H. Graphene Oxide-Silver Nanoparticles Nanocomposite Stimulates Differentiation in Human Neuroblastoma Cancer Cells (SH-SY5Y). *Int. J. Mol. Sci.* **2017**, *18*, 2549. [CrossRef]
31. Ferrari, A.C. Raman spectroscopy of graphene and graphite: Disorder, electron–phonon coupling, doping and nonadiabatic effects. *Solid State Commun.* **2007**, *143*, 47–57. [CrossRef]
32. Cho, E.S.; Ruminski, A.M.; Aloni, S.; Liu, Y.-S.; Guo, J.; Urban, J.J. Graphene oxide/metal nanocrystal multilaminates as the atomic limit for safe and selective hydrogen storage. *Nat. Commun.* **2016**, *7*, 10804. [CrossRef]
33. Fiorillo, M.; Verre, A.F.; Iliut, M.; Peiris-Pagés, M.; Ozsvari, B.; Gandara, R.; Cappello, A.R.; Sotgia, F.; Vijayaraghavan, A.; Lisanti, M.P. Graphene oxide selectively targets cancer stem cells, across multiple tumor types: Implications for non-toxic cancer treatment, via "differentiation-based nano-therapy". *Oncotarget* **2015**, *6*, 3553. [CrossRef]
34. Liao, K.-H.; Lin, Y.-S.; Macosko, C.W.; Haynes, C.L. Cytotoxicity of graphene oxide and graphene in human erythrocytes and skin fibroblasts. *ACS Appl. Mater. Interfaces* **2011**, *3*, 2607–2615. [CrossRef]
35. Choi, Y.-J.; Gurunathan, S.; Kim, J.-H. Graphene Oxide–Silver Nanocomposite Enhances Cytotoxic and Apoptotic Potential of Salinomycin in Human Ovarian Cancer Stem Cells (OvCSCs): A Novel Approach for Cancer Therapy. *Int. J. Mol. Sci.* **2018**, *19*, 710. [CrossRef]
36. Gurunathan, S.; Kim, J.-H. Biocompatible gold nanoparticles ameliorate retinoic acid-induced cell death and induce differentiation in f9 teratocarcinoma stem cells. *Nanomaterials* **2018**, *8*, 396. [CrossRef]
37. Lammel, T.; Boisseaux, P.; Fernández-Cruz, M.-L.; Navas, J.M. Internalization and cytotoxicity of graphene oxide and carboxyl graphene nanoplatelets in the human hepatocellular carcinoma cell line Hep G2. *Part. Fibre Toxicol.* **2013**, *10*, 27.
38. Xu, M.; Zhu, J.; Wang, F.; Xiong, Y.; Wu, Y.; Wang, Q.; Weng, J.; Zhang, Z.; Chen, W.; Liu, S. Improved in vitro and in vivo biocompatibility of graphene oxide through surface modification: Poly (acrylic acid)-functionalization is superior to PEGylation. *ACS Nano* **2016**, *10*, 3267–3281. [CrossRef]
39. Li, Y.; Wu, Q.; Zhao, Y.; Bai, Y.; Chen, P.; Xia, T.; Wang, D. Response of microRNAs to in vitro treatment with graphene oxide. *ACS Nano* **2014**, *8*, 2100–2110. [CrossRef]
40. Gurunathan, S.; Han, J.W.; Eppakayala, V.; Jeyaraj, M.; Kim, J.-H. Cytotoxicity of biologically synthesized silver nanoparticles in MDA-MB-231 human breast cancer cells. *Biomed. Res. Int.* **2013**, 535796. [CrossRef]
41. Chatterjee, N.; Eom, H.-J.; Choi, J. A systems toxicology approach to the surface functionality control of graphene–cell interactions. *Biomaterials* **2014**, *35*, 1109–1127. [CrossRef]
42. Cho, Y.C.; Pak, P.J.; Joo, Y.H.; Lee, H.-S.; Chung, N. In vitro and in vivo comparison of the immunotoxicity of single-and multi-layered graphene oxides with or without pluronic F-127. *Sci. Rep.* **2016**, *6*, 38884. [CrossRef]
43. Qin, Y.; Zhou, Z.-W.; Pan, S.-T.; He, Z.-X.; Zhang, X.; Qiu, J.-X.; Duan, W.; Yang, T.; Zhou, S.-F. Graphene quantum dots induce apoptosis, autophagy, and inflammatory response via p38 mitogen-activated protein kinase and nuclear factor-κB mediated signaling pathways in activated THP-1 macrophages. *Toxicology* **2015**, *327*, 62–76. [CrossRef] [PubMed]
44. Pelin, M.; Sosa, S.; Prato, M.; Tubaro, A. Occupational exposure to graphene based nanomaterials: Risk assessment. *Nanoscale* **2018**, *10*, 15894–15903. [CrossRef]

45. Tabish, T.A.; Pranjol, M.Z.I.; Jabeen, F.; Abdullah, T.; Latif, A.; Khalid, A.; Ali, M.; Hayat, H.; Winyard, P.G.; Whatmore, J.L. Investigation into the toxic effects of graphene nanopores on lung cancer cells and biological tissues. *Appl. Mater. Today* **2018**, *12*, 389–401. [CrossRef]
46. Valavanidis, A.; Vlachogianni, T.; Fiotakis, C. 8-hydroxy-2′-deoxyguanosine (8-OHdG): A critical biomarker of oxidative stress and carcinogenesis. *J. Environ. Sci. Health Part C* **2009**, *27*, 120–139. [CrossRef]
47. Liu, Y.; Luo, Y.; Wu, J.; Wang, Y.; Yang, X.; Yang, R.; Wang, B.; Yang, J.; Zhang, N. Graphene oxide can induce in vitro and in vivo mutagenesis. *Sci. Rep.* **2013**, *3*, 3469. [CrossRef]
48. Gurunathan, S.; Han, J.W.; Eppakayala, V.; Kim, J.-H. Green synthesis of graphene and its cytotoxic effects in human breast cancer cells. *Int. J. Nanomed.* **2013**, *8*, 1015. [CrossRef]
49. Gurunathan, S.; Han, J.W.; Kim, E.S.; Park, J.H.; Kim, J.-H. Reduction of graphene oxide by resveratrol: A novel and simple biological method for the synthesis of an effective anticancer nanotherapeutic molecule. *Int. J. Nanomed.* **2015**, *10*, 2951. [CrossRef]
50. Zhi, X.; Fang, H.; Bao, C.; Shen, G.; Zhang, J.; Wang, K.; Guo, S.; Wan, T.; Cui, D. The immunotoxicity of graphene oxides and the effect of PVP-coating. *Biomaterials* **2013**, *34*, 5254–5261. [CrossRef]
51. Ma, J.; Liu, R.; Wang, X.; Liu, Q.; Chen, Y.; Valle, R.P.; Zuo, Y.Y.; Xia, T.; Liu, S. Crucial role of lateral size for graphene oxide in activating macrophages and stimulating pro-inflammatory responses in cells and animals. *ACS Nano* **2015**, *9*, 10498–10515. [CrossRef]
52. Redza-Dutordoir, M.; Averill-Bates, D.A. Activation of apoptosis signalling pathways by reactive oxygen species. *Biochim. Biophys. Acta (BBA) Mol. Cell Res.* **2016**, *1863*, 2977–2992. [CrossRef]
53. Yuan, Y.-G.; Gurunathan, S. Combination of graphene oxide–silver nanoparticle nanocomposites and cisplatin enhances apoptosis and autophagy in human cervical cancer cells. *Int. J. Nanomed.* **2017**, *12*, 6537. [CrossRef]
54. Yoshida, K.; Miki, Y. The cell death machinery governed by the p53 tumor suppressor in response to DNA damage. *Cancer Sci.* **2010**, *101*, 831–835. [CrossRef]
55. Cheng, C.Y.; Wong, E.W.; Lie, P.P.; Li, M.W.; Su, L.; Siu, E.R.; Yan, H.H.; Mannu, J.; Mathur, P.P.; Bonanomi, M. Environmental toxicants and male reproductive function. *Spermatogenesis* **2011**, *1*, 2–13. [CrossRef] [PubMed]
56. Hechelhammer, L.; Störkel, S.; Odermatt, B.; Heitz, P.U.; Jochum, W. Epidermal growth factor receptor is a marker for syncytiotrophoblastic cells in testicular germ cell tumors. *Virchows Arch.* **2003**, *443*, 28–31. [CrossRef]
57. Porta, C.; Paglino, C.; Mosca, A. Targeting PI3K/Akt/mTOR signaling in cancer. *Front. Oncol.* **2014**, *4*, 64. [CrossRef]
58. Song, G.; Ouyang, G.; Bao, S. The activation of Akt/PKB signaling pathway and cell survival. *J. Cell. Mol. Med.* **2005**, *9*, 59–71. [CrossRef]
59. Pei, S.; Zhao, J.; Du, J.; Ren, W.; Cheng, H.-M. Direct reduction of graphene oxide films into highly conductive and flexible graphene films by hydrohalic acids. *Carbon* **2010**, *48*, 4466–4474. [CrossRef]
60. Qi, X.; Zhou, T.; Deng, S.; Zong, G.; Yao, X.; Fu, Q. Size-specified graphene oxide sheets: Ultrasonication assisted preparation and characterization. *J. Mater. Sci.* **2014**, *49*, 1785–1793. [CrossRef]
61. Liu, X.; Gan, W.; Zou, Y.; Yang, B.; Su, Z.; Deng, J.; Wang, L.; Cai, J. Elevated levels of urinary markers of oxidative DNA and RNA damage in type 2 diabetes with complications. *Oxidative Med. Cell. Longev.* **2016**, 4323198. [CrossRef]
62. Gurunathan, S.; Qasim, M.; Park, C.; Yoo, H.; Kim, J.-H.; Hong, K. Cytotoxic Potential and Molecular Pathway Analysis of Silver Nanoparticles in Human Colon Cancer Cells HCT116. *Int. J. Mol. Sci.* **2018**, *19*, 2269. [CrossRef] [PubMed]

nanomaterials

MDPI

Article

Improving of the Photovoltaic Characteristics of Dye-Sensitized Solar Cells Using a Photoelectrode with Electrospun Porous TiO_2 Nanofibers

Min Su Jo [1], Jung Sang Cho [1,*], Xuan Liang Wang [2], En Mei Jin [2,*], Sang Mun Jeong [2,*] and Dong-Won Kang [3]

1 Department of Engineering Chemistry, Chungbuk National University, 1 Chungdae-ro, Seowon-Gu, Cheongju, Chungbuk 28644, Korea; jms039@naver.com
2 Department of Chemical Engineering, Chungbuk National University, 1 Chungdae-ro, Seowon-Gu, Cheongju, Chungbuk 28644, Korea; wangxuanleon@gmail.com
3 School of Energy Systems Engineering, Chung-Ang University, Seoul 06974, Korea; kangdwn@cau.ac.kr
* Correspondence: jscho@chungbuk.ac.kr (J.S.C.); jinenmei@chungbuk.ac.kr (E.M.J.); smjeong@chungbuk.ac.kr (S.M.J.); Tel.: +82-43-261-2489 (J.S.C.); +82-70-8667-0097 (E.M.J.); +82-43-261-3369 (S.M.J.)

Received: 11 December 2018; Accepted: 8 January 2019; Published: 12 January 2019

Abstract: Porous TiO_2 nanofibers (PTFs) and dense TiO_2 nanofibers (DTFs) were prepared using simple electrospinning for application in dye-sensitized solar cells (DSSCs). TiO_2 nanoparticles (TNPs) were prepared using a hydrothermal reaction. The as-prepared PTFs and DTFs (with a fiber diameter of around 200 nm) were mixed with TNPs such as TNP-PTF and TNP-DTF nanocomposites used in photoelectrode materials or were coated as light scattering layers on the photoelectrodes to improve the charge transfer ability and light harvesting effect of the DSSCs. The as-prepared TNPs showed a pure anatase phase, while the PTFs and DTFs showed both the anatase and rutile phases. The TNP-PTF composite (TNP:PTF = 9:1 wt.%) exhibited an enhanced short circuit photocurrent density (J_{sc}) of 14.95 ± 1.03 mA cm^{-2} and a photoelectric conversion efficiency (PCE, η) of 5.4 ± 0.17% because of the improved charge transport and accessibility for the electrolyte ions. In addition, the TNP/PTF photoelectrode showed excellent light absorption in the visible region because of the mountainous nature of light induced by the PTF light scattering layer. The TNP/PTF photoelectrode showed the highest J_{sc} (16.96 ± 0.79 mA cm^{-2}), η (5.9 ± 0.13%), and open circuit voltage (V_{oc}, 0.66 ± 0.02 V).

Keywords: electrospinning; nanocomposites; porous TiO_2 nanofiber; light harvesting; additive; dye-sensitized solar cells

1. Introduction

Dye-sensitized solar cells (DSSCs) are basically thin-layer solar cells consisting of two sandwich-type transparent conduction oxide (TCO) electrodes. One electrode is a highly colored photoelectrode with a few micron-thick layers of mesoporous TiO_2 or other semiconductors (ZnO, SnO_2, and Nb_2O_5) coated with a photosensitizer, while the other is a Pt-based counter-electrode [1–5]. The space between the two electrodes is filled with an organic electrolyte containing a redox mediator (I^-/I_3^-), usually a mixture of iodine and iodide in organic solvents such as acetonitrile [6–10]. DSSCs, which are cost-effective and have high theoretical efficiencies, are available in various colors depending on the dye. DSSCs co-photosensitized with an organic dye of alkoxysilyl-anchor dye (ADEKA-1, molecular structures of carbazole dyes with a trimethoxysilyl group) and a carboxy-anchor organic dye of LEG4 anchored to nanocrystalline TiO_2 films show a photoelectric conversion efficiency (PCE) of up to 14.7% (under 1 sun illumination) [11]. The theoretical PCE limit of DSSCs using a simple junction configuration (under standard test conditions) is reported to be 33% [12,13]. However, the PCE of present DSSCs is significantly lower than their theoretical efficiency. The PCE response of DSSCs depends on their open circuit voltage (V_{oc}, V), photocurrent density (J_{sc}, mA cm^{-2}), and fill factor (*FF*, %) [14,15]. The V_{oc} is estimated from the difference between the quasi-Fermi levels of semiconductor materials, such as TiO_2, and the redox potential of the electrolyte (I^-/I_3^-). Therefore, the V_{oc} of DSSCs is closely related to the properties of the semiconductor materials constituting them. The J_{sc} and *FF* of DSSCs depend on the adsorption of the dye and the charge transfer in the photoelectrode [15–17].

TiO_2 typically used for a photoelectrode is chemically stable, non-toxic, and readily available in vast quantities [18–22]. For the enhanced photovoltaic properties of TiO_2 for DSSC, it is imperative to improve its surface area, dye molecular adsorption activity, light harvesting effect, and charge transport. There have been many reports on the use of metal oxide additives, such as SiO_2, SnO_2, and ZrO_2, for improving the photovoltaic properties of TiO_2 nanoparticles [23–25]. In our previous study [26], SnO_2 hollow fibers were used as the additives for the TiO_2 photoelectrodes in order to improve the electron transport of DSSCs. The addition of SnO_2 hollow fibers accelerated the electron transfer and improved the electrochemical properties of the DSSC. The PCE of the DSSC with the SnO_2 hollow fiber-added TiO_2 photoelectrode was approximately 11% higher (5.43%) than that of the DSSC with pure TiO_2 photoelectrode (4.89%). Swathy et al. [27] used nanostructured (≈400 nm particle size) anatase titania spheres as the light scattering layer for DSSCs. These DSSCs showed about two times higher J_{sc} (10.44 mA cm^{-2}) and two times higher PCE (4.92%) than those of TiO_2 photoelectrodes without scattering layers (J_{sc} of 5.83 mA cm^{-2}, PCE of 2.7%). The photovoltaic performance of the DSSCs was measured under an AM1.5G solar spectrum with a light intensity of 100 mW cm^{-2}.

In summary, the PCE of DSSCs can be easily improved by altering the design of their photoelectrodes. In this study, we prepared seven TiO_2 photoelectrodes to improve the PCE of DSSCs, as shown in Scheme 1. The TiO_2 photoelectrodes were fabricated using TiO_2 nanoparticles and the porous TiO_2 nanofibers (PTFs), and dense TiO_2 nanofibers (DTFs) were used as additives or light scattering layer. The PTFs and DTFs were prepared using simple electrospinning. TiO_2 nanoparticles (TNPs) were prepared using a hydrothermal reaction and were used for fabricating the photoelectrodes. The as-prepared PTFs and DTFs were used as additives or were coated as light scattering layers to improve the charge transfer ability and light harvesting effect of the DSSCs.

2. Materials and Methods

2.1. Preparation of Porous TiO_2 Nanofibers (PTFs)

The porous TiO_2 nanofibers (PTF) were prepared via an electrospinning process and subsequent heat treatment. The spinning solution was prepared by adding 2 mL butyl titanate (TBT, $C_{16}H_{36}O_4Ti$, ≥97%, Kanto Chemical Co. Inc, Tokyo, Japan), 1 g polyacrylonitrile (PAN, Mw: 150,000, Sigma-Aldrich, St. Louis, MO, USA), 4 g polystyrene (PS, Mw: ≈192,000, Sigma-Aldrich), and 2 mL of acetic acid (99.7%, Daejung Chemicals and Metals, Siheung, Korea) in 40 mL of *N*,*N*-dimethylformamide (DMF, ≥99.5%, Samchun Chemical, Seoul, Korea). The resulting solution was stirred at room temperature for 24 h to ensure complete dissolution and was then loaded into a plastic syringe equipped with a 25 G stainless steel needle ($Ø_{in}$ = 0.25 mm, $Ø_{out}$ = 0.51 mm). The needle was connected to a high-voltage supply and a voltage of 20 kV was applied between the needle and the collector. The distance between the needle tip and the drum collector was set at 15 cm at a flow rate of 2 mL h^{-1}. The rotation of the drum was maintained at 150 rpm. The resulting electrospun fibers were heated at 100 °C in air for 6 h to remove the solvent, and then heat-treated at 500 °C for 3 h (heating rate was 5 °C min^{-1}) in air to remove the polymer.

2.2. Preparation of Dense TiO_2 Nanofibers (DTFs)

The dense TiO_2 nanofibers (DTF) were also prepared via an electrospinning process and subsequent heat treatment. The spinning solution was prepared by adding 4 mL TBT, 2 g polyvinylpyrrolidone (PVP, M.W. 1,300,000, Alfa Aesar, Haverhill, MA, USA), and 8 mL of acetic acid in 20 mL of ethanol (99.7%, Daejung Chemicals and Metals, Siheung, Korea). The resulting solution was electrospun and heat-treated under the same condition as described above for PTF.

2.3. Preparation of Anatage TiO_2 Nanoparticles (TNPs)

TNPs were synthesized via the hydrothermal method. In a typical reaction, 15 mL of titanium tetraisopropoxide (TTIP, 99.9%, Aldrich) was dissolved in 50 mL of distilled water and the resulting solution was vigorously stirred for a few minutes before adding 0.7 mL of ammonia solution (28–30%, SAMCHUN) to it. This solution was magnetically stirred at 350 rpm for 3 h to obtain a homogeneous mixture. The reaction mixture so obtained was transferred to a Teflon-lined autoclave and was then heated at 200 °C for 5 h. Finally, the white precipitate was washed with distilled water and ethanol and then dried at 80 °C for 12 h.

2.4. Preparation of DSSCs

TiO_2 photoelectrodes were fabricated using TNP or TNP-DTF or TNP-PTF nanocomposites. The TiO_2 pastes were prepared by mixing acetylacetone (99%, Sigma-Aldrich), hydroxypropyl cellulose (99%, Aldrich), and distilled water. Each paste was coated on a fluorine-doped tin oxide (8 Ω/sq, Pilkington) substrate, which was subsequently sintered at 450 °C for 30 min. The obtained TiO_2 films were immersed in a 0.5 mM ethanol solution of N719 dye (bis-(isothiocyanato)bis (2,2-bipyridyl-4,4-dicarboxylato)ruthenium(II)bis-tetrabutylammonium, Solaronix) for 5 h. The active area of the photoelectrode was 0.25 cm^2. The sandwich-type DSSC was assembled using a Pt counter-electrode and a dye-sensitized photoelectrode. A redox I^-/I_3^- electrolyte was introduced between the electrodes. The preparation methods of the counter electrode and I^-/I_3^- electrolyte were described in detail in our previous report [17].

In this study, we prepared seven kinds of photoelectrodes, as shown in Scheme 1. The photoelectrodes with only TNP, PTFs, and DTFs were labeled as TNP, PTF, and DTF, respectively. The TNP photoelectrodes with 10 wt.% of PTFs and DTFs were denoted as TNP-PTF and TNP-DTF, respectively. The photoelectrodes with PTF and DTF light scattering layers were denoted as TNP/PTF and TNP/DTF, respectively.

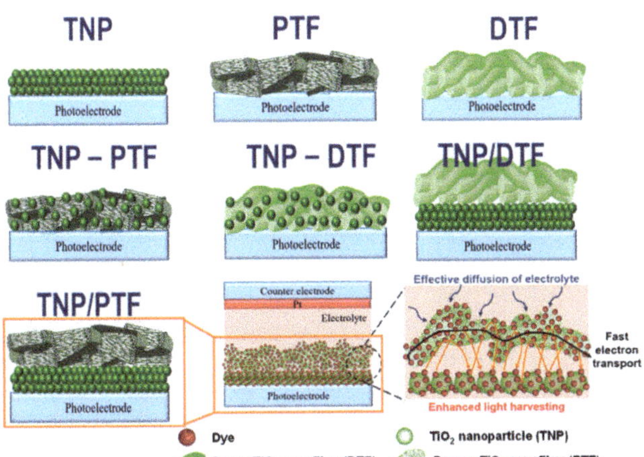

Scheme 1. Various photoelectrodes for DSSCs.

2.5. Characteristic and Measurements

The crystalline phases and morphologies of the prepared materials were identified using X-ray diffraction (XRD, Ultima IV, Rigaku, Japan), field-emission scanning electron microscopy (FE-SEM; LEO-1530, Carl Zeiss, Oberkochen, Germany), and field-emission transmission electron microscopy (FE-TEM, 200KV, JEM 2100F, JEOL, Tokyo, Japan). To estimate the specific surface area, pore volume, and pore size distribution of the materials, and their N_2 adsorption/desorption isotherms were measured at 77 K using a surface area analyzer (ASAP2020, Micromeritics, Norcross, GA, USA). The Brunauer–Emmett–Teller (BET) method and Barrett–Joyner–Halenda (BJH) model were used to estimate the specific surface area and pore size distribution of the samples, respectively. Ultraviolet-visible (UV-Vis-NIR) reflectance spectra were obtained using a UV-Vis-NIR spectrophotometer (Cary 5G, Varian, Palo Alto, CA, USA) equipped with diffuse reflectance accessory (integral sphere). Total reflectance spectrum (Total reflectance = Diffuse reflectance + Specular reflectance) was recorded in the spectral range of 200 to 800 nm at a scan rate of 600 nm min^{-1}. In addition, the UV-Vis spectrum of N719 solution was measured using S-3100 equipment (Sinco, Seoul, Korea). The photovoltaic properties of the DSSCs were evaluated by recording their current density–voltage characteristics under illumination from a Polaronix K201 (McScience, Suwon, Korea) equipped with a K401 CW150 lamp power supply and an AM 1.5G filter (100 mW cm^{-2}).

3. Results

The morphological features of PTFs, DTFs, and TNPs were observed by FE-SEM and FE-TEM, as shown in Figure 1. Figure 1a shows PTF prepared using a spinning process, in which numerous pores in the structure were confirmed from the fiber surface and cross-sectioned inset image. The PS in the spinning solution was separated with PAN during electrospinning. Subsequently, PS in the as-spun fibers was decomposed into CO_2 gas during heat treatment, which formed the numerous pores between TiO_2 crystals in the structure. On the other hand, DTF fibers prepared from the solution without PS showed dense structure in Figure 1b. Without the effect of phase separation by adding PS, pores were not generated in the structure. Therefore, it was hard to observe any pores, even in the cross-sectioned inset image in Figure 1b. The mean fiber diameter of PTF and DTF measured from FE-TEM images were both 200 nm. From the elemental mapping images in Figure 1d,e, both TiO_2 nanofibers showed uniformly distributed Ti and O elements over the nanostructure, but the C element was negligible, which proves the completely decomposition of C during heat treatment. The TNPs were rice-shaped and had an average particle size of about 30 nm (Figure 1c). Figure 1d–f shows the FETEM and EDX mapping results of the PTF, DTF, and TNP powders. The EDX mapping results show that only Ti and O were present in the samples. C was detected on the carbon tape used for the sampling. The EDX results also showed that the PTFs were less dense as compared to the DTFs. BET surface area measurements were carried out to determine the pore diameters and surface areas of the samples quantitatively.

The pore diameter and surface areas of the PTFs, DTFs, and TNPs were measured by carrying out BET measurements. The pore distribution and diameter determine the fast charge transfer and light harvesting effect of porous fibers. However, a fiber structure is not good for the adsorption of light-absorbing dyes. Adsorption of dye molecules is closely related to the specific surface area of photoelectrode materials. Hence, in this study, we fabricated various photoelectrodes using the TNPs, PTFs, and DTFs. The TNP, PTF, DTF, TNP-PTF, TNP-DTF, TNP/PTF, and TNP/DTF photoelectrodes were used to assemble DSSCs. The photovoltaic properties of these DSSCs were investigated. Figure 2 shows the cross-section and top view of the TNP/PTF, TNP/DTF, and TNP photoelectrodes. The thickness of the TNP photoelectrode was about 8 ± 0.6 μm and the PTF and DTF light scattering layers were coated on it. The top-view of TNP/PTF and TNP/DTF (Figure 2c,d) revealed that TNP/PTF showed porosity and a loofah-like formation. These pores could improve the light-harvesting and electrolyte diffusion of the photoelectrode and hence improved the charge transport and accessibility of the electrolyte ions [28–31].

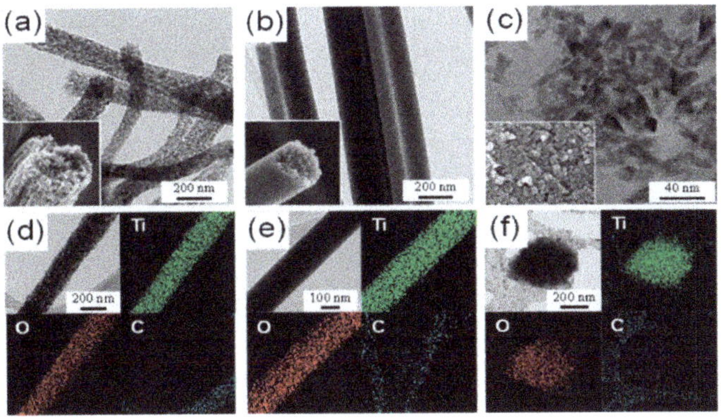

Figure 1. FESEM and FETEM images of the (**a**) PTF, (**b**) DTF, and (**c**) TNP powders. FETEM and EDX mapping of the (**d**) PTF, (**e**) DTF, and (**f**) TNP powders.

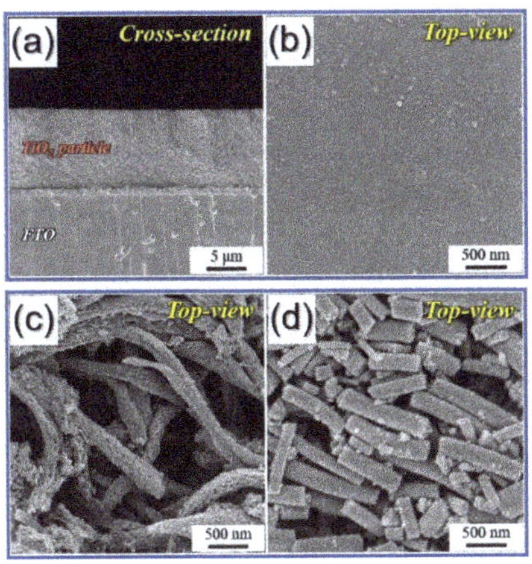

Figure 2. FESEM images of the (**a**) cross-section view of the TNP photoelectrode, and top-view of the (**b**) TNP, (**c**) TNP/PTF and (**d**) TNP/DTF photoelectrodes.

N_2 adsorption of all the samples was carried out and the results are shown in Figure 3 and Table 1. The TNPs exhibited a specific surface area of 140 m^2 g^{-1}. The surface areas of the PTFs and DTFs were lower than that of the TNPs. The TNPs, PTFs, and DTPs showed type-IV adsorption isotherms, indicating the presence of mesoporous structures with an average pore size of 7.13, 18.47, and 5.48 nm, respectively. In the case of the TNPs, the hysteresis loop was more prominent for the relative pressures (P/P_0) in the range of 0.7–1.0, indicating the presence of many pores. Figure 3b shows the pore diameter distribution of the samples (within 0–100 nm). The PTFs showed a large mesopore volume of 10–50 nm. They exhibited a hierarchical pore structure consisting mainly of mesopores along with some micropores. Therefore, the use of PTF photoelectrodes can improve the electrode-electrolyte interactions and shorten the ionic transport paths, thus promoting the charge transport and enhancing the accessibility for electrolyte ions.

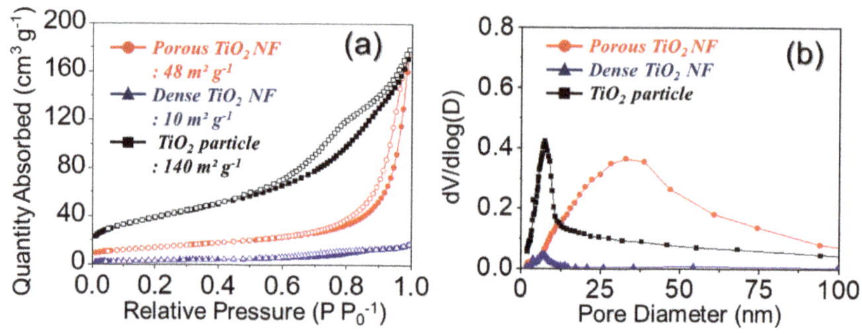

Figure 3. (**a**) N_2 adsorption isotherms and (**b**) pore-size distribution of the PTF, DTF, and TNP powders.

Table 1. BET and crystal structure of the TNP, PTF, and DTF powders and the photovoltaic parameters of different photoelectrode used DSSC.

Photo-Electrodes	BET (m^2 g^{-1})	Pore Size (nm)	Pore Volume (cm^3 g^{-1})	Crystal Structure	V_{oc} (V)	J_{sc} (mA cm^{-2})	FF (%)	η (%)
TNP	140.08	7.13	0.2763	Anatase	0.69 ± 0.02	12.51 ± 0.60	53 ± 2	4.6 ± 0.07
PTF	48.01	18.47	0.2794	Anatase, Rutile	0.65 ± 0.01	10.30 ± 0.96	56 ± 2	3.8 ± 0.19
DTF	23.16	5.48	0.0460	Anatase, Rutile	0.66 ± 0.01	10.18 ± 1.24	48 ± 7	3.2 ± 0.15
TNP-PTF	-	-	-	-	0.68 ± 0.01	14.95 ± 1.03	54 ± 2	5.4 ± 0.17
TNP-DTF	-	-	-	-	0.68 ± 0.01	13.42 ± 0.50	55 ± 1	5.1 ± 0.16
TNP/PTF	-	-	-	-	0.66 ± 0.02	16.96 ± 0.79	56 ± 2	5.9 ± 0.13
TNP/DTF	-	-	-	-	0.66 ± 0.01	14.39 ± 0.50	54 ± 1	5.2 ± 0.13

The structural properties of the PTF, DTF, and TNP powders were investigated using their XRD patterns shown in Figure 4. The PTFs and DTFs showed mixed anatase (space group *I41/amd*, card no. 21-1272 in the JCPDS database) and rutile (card no. 21-1276 in the JCPDS database) phases. On the other hand, the TNPs showed only the anatase phase [32,33]. The dark and white squares in Figure 4 correspond to the anatase and rutile phases, respectively. The XRD results showed that the rutile phase fraction of the DTFs (rutile phase peak intensity was higher) was higher than that of the PTF. The rutile phase has a slightly lower bandgap than the anatase phase and can absorb a little sunlight in the near-UV region. The open-circuit voltage of DSSCs can be controlled by adjusting the position of the TiO_2 conduction band. Most of the studies in this context have focused on anatase TiO_2. However, rutile TiO_2 is easy to produce and has superior light-scattering properties, which is beneficial for effective light-harvesting [34]. Therefore, we used the PTFs and DTFs (which consisted of the rutile phase) as light scattering layers for DSSCs.

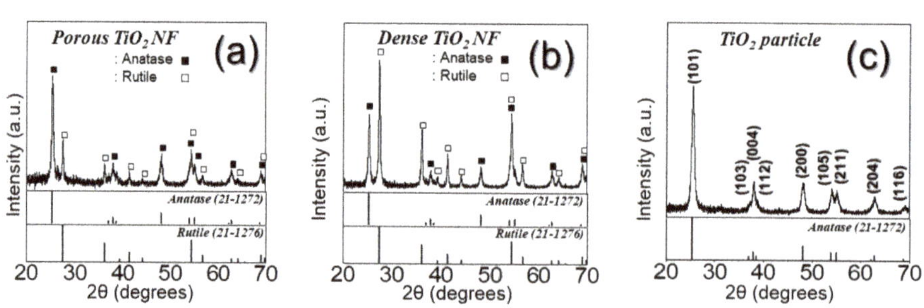

Figure 4. X-ray diffraction of the (**a**) PTF, (**b**) DTF, and (**c**) TNP powders.

Figure 5 shows the photocurrent density–voltage curves of the DSSCs fabricated using different photoelectrodes. As shown in Figure 5, the TNP photoelectrode showed a higher open-circuit voltage

(V_{oc}) (0.69 ± 0.02 V) than the PTF and DTF photoelectrodes. The V_{oc} of DSSCs can be controlled by adjusting the position of the TiO_2 conduction band. Hence, the V_{oc} of the PTF and DTF photoelectrodes (having the rutile crystalline structure) was lower than that of the anatase TNP photoelectrode. In addition, the photocurrent densities of the PTF and DTF photoelectrodes were lower than that of the TNP photoelectrode because of the loading of a small amount of dye. Dye loading of a photoelectrode is closely related to the specific surface area of TiO_2; the larger the specific surface area of TiO_2, the larger is the amount of dye that can be adsorbed on the TiO_2 surface. The addition of the PTFs and DTFs to the TNP photoelectrodes resulted in an increase in the short circuit photocurrent density (J_{sc}) and PCE (η) because of the improved charge transport and enhanced accessibility for the electrolyte ions. On the other hands in order to improve the light harvesting effect of the DSSCs, the PTF or DTF light scattering layers were coated on the TNP photoelectrodes. As the result, the TNP/PTF photoelectrode showed the highest J_{sc} and η. The high J_{sc} can be attributed to the increased light harvesting effect due to the PTF light-scattering layer. The rutile phase-mixed PTF acted as an energy barrier and increased the physical separation between the injected electrons and oxidized dyes/redox couples, thereby retarding the recombination reactions in the resulting DSSCs [35]. The detailed photocurrent density–voltage results are given in Table 1. The TNP/PTF photoelectrode-based DSSC showed the highest η of 5.9 ± 0.13%, which is approximately 28% higher than that of the TNP photoelectrode-based DSSCs (4.6 ± 0.07%).

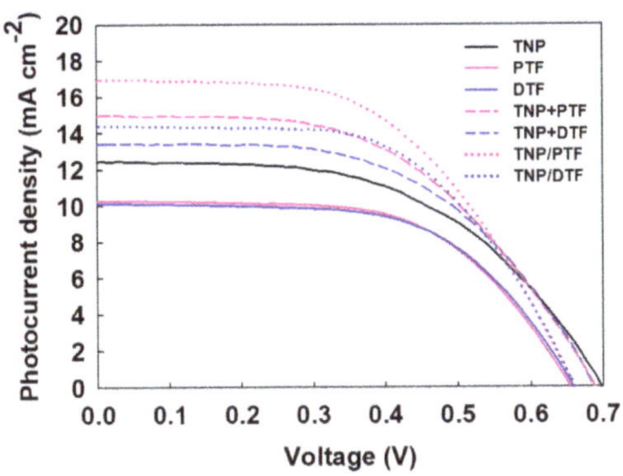

Figure 5. Photocurrent density–voltage curves of the DSSCs with different photoelectrodes.

The UV-Vis absorption spectra of the TNP, TNP/DTF, TNP/PTF photoelectrodes, and N719 dye are shown in Figure 6. As shown in Figure 6a, all the samples showed strong absorption in the UV region (wavelengths lower than 250 nm) before dye adsorption. It can be observed that the absorption edges of the TNP/DTF and TNP/PTF photoelectrodes slightly red-shifted with respect to the TNP photoelectrode, suggesting that the TNP/DTF and TNP/PTF photoelectrodes caused a small decrease in the energy band gap. In addition, the TNP/DTF and TNP/PTF photoelectrodes showed an absorption peak in the near-UV region at 330 nm, which can be attributed to their rutile structure. All the samples showed almost no absorption in the visible region. As shown in Figure 6b, the N719 dye showed an absorption peak in the visible region at 530 nm. The intensity of this peak was high in the case of the TNP/PTF photoelectrode. This indicates that the use of the PTF light scattering layer (TNP/PTF) coating increased the visible region absorption of the DSSC.

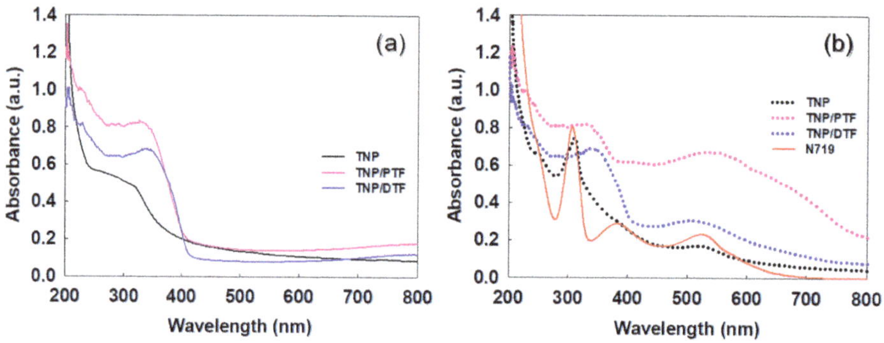

Figure 6. UV-Vis absorption spectra of the different photoelectrodes (**a**) before and (**b**) after N719 dye adsorption.

4. Conclusions

Porous and dense TiO_2 nanofibers (200 nm fiber diameter) were successfully prepared using electrospinning and were used as additives or light-scattering layers to enhance the photovoltaic properties of DSSC. Both the PTFs and DTFs showed mixed phases of anatase and rutile TiO_2 crystals. The addition of 10 wt.% of PTFs or DTFs increased the short circuit photocurrent density (J_{sc}) of the TNP photoelectrode because of an improvement in the charge transport and accessibility for the electrolyte ions. In addition, the TNP/PTF photoelectrode improved the visible light absorption of the DSSC because of the mountainous nature of light due to the PTF light scattering layer. The highest short circuit photocurrent density was shown by the TNP/PTF photoelectrode (16.96 $\pm$ 0.79 mA cm^{-2}). The TNP/PTF photoelectrode showed a high PCE of 5.9 $\pm$ 0.13% and an open-circuit voltage (V_{oc}) of 0.66 $\pm$ 0.02 V. These results show that the use of scattering layers made up of PTFs is an effective approach to improve the photovoltaic performance of DSSCs.

Author Contributions: J.S.C. and S.M.J. conceived the overall approach and supervised the project. M.S.J. and X.L.W. performed experiments and E.M.J. designed experiments, analyzed the data, and wrote this manuscript, which was corrected by J.S.C., S.M.J., and D.-W.K.

Funding: Please add: This research was supported by Basic Science Research Program through the National Research Foundation of Korea (NRF) funded by the Korea government (2017R1D1A1B03031989; 2017M1A2A2087577; 2018R1A4A1024691).

Conflicts of Interest: The authors declare no conflict of interest.

References

1. Tehare, K.K.; Navale, S.T.; Stadler, F.J.; He, Z.; Yang, H.; Xiong, X.; Liu, X.; Mane, R.S. Enhanced DSSCs Performance of TiO_2 Nanostructure by Surface Passivation Layers. *Mater. Res. Bull.* **2018**, *99*, 491–495. [CrossRef]
2. Giovannetti, R.; Zannotti, M.; Alibabaei, L.; Ferraro, S. Equilibrium and Kinetic Aspects in the Sensitization of Monolayer Transparent TiO_2 Thin Films with Porphyrin Dyes for DSSC Applications. *Int. J. Photoenergy* **2014**, *2014*, 1–9. [CrossRef]
3. Giannouli, M.; Govatsi, K.; Syrrokostas, G.; Yannopoulos, S.N.; Leftheriotis, G. Factors Affecting the Power Conversion Efficiency in ZnO DSSCs: *Nanowire vs. Nanoparticles*. *Materials* **2018**, *11*, 411. [CrossRef] [PubMed]
4. Kiran, S.; Naveen Kumar, S.K. Preparation and Thickness Optimization of TiO_2/Nb_2O_5 Photoanode for Dye Sensitized Solar Cells. *Mater. Today* **2018**, *5*, 10797–10804. [CrossRef]
5. Sasikumar, R.; Chen, T.-W.; Chen, S.M.; Rwei, S.-P.; Ramaraj, S.K. Developing the Photovoltaic Performance of Dye-sensitized Solar Cells (DSSCs) Using a SnO_2-Doped Graphene Oxide Hybrid Nanocomposite as a Photo-Anode. *Opt. Mater.* **2018**, *79*, 345–352. [CrossRef]
6. Jin, E.M.; Jeong, S.M.; Kang, H.-C.; Gu, H.-B. Photovoltaic Effect of Metal-Doped TiO_2 Nanoparticles for Dye-Sensitized Solar Cells. *ECS J. Solid State Sci. Technol.* **2016**, *5*, Q109–Q114. [CrossRef]

7. Kim, D.-W.; Jeong, Y.-B.; Kim, S.-H.; Lee, D.-Y.; Song, J.-S. Photovoltaic Performance of Dye-Sensitized Solar Cell Assembled with Gel Polymer Electrolyte. *J. Power Sources* **2005**, *149*, 112–116. [CrossRef]
8. Palomares, E.; Clifford, J.N.; Haque, S.A.; Lutz, T.; Durrant, J.R. Slow Charge Recombination in Dye-Sensitized Solar Cells (DSSC) Using Al_2O_3 Coated Nanoporous TiO_2 Films. *Chem. Commun.* **2002**, *14*, 1464–1465. [CrossRef]
9. Bhattacharya, S.; Pal, A.; Jana, A.; Datta, J. Synthesis and Characterization of CdS Canoparticles Decorated TiO_2 Matrix for an Efficient N3 Based Dye Sensitized Solar Cell (DSSC). *J. Mater. Sco. Mater. Electron.* **2016**, *27*, 12438–12445. [CrossRef]
10. Yue, G.; Wu, J.; Xiao, Y.; Lin, J.; Huang, M.; Fan, L.; Yao, Y. A Dye-Sensitized Solar Cell Based on PEDOT: PSS Counter Electrode. *Chin. Sci. Bull.* **2013**, *58*, 559–566. [CrossRef]
11. Kakiage, K.; Aoyama, Y.; Yano, T.; Oya, K.; Fujisawa, J.-I.; Hanaya, M. Highly-Efficient Dye-Sensitized Solar Cells With Collaborative Sensitization by Silyl-Anchor and Carboxyl-Anchor Dyes. *Chem. Commun.* **2015**, *51*, 15894–15897. [CrossRef] [PubMed]
12. Lee, Y.; Kang, M. The Optical Properties of Nanoporous Structured Titanium Dioxide and the Photovoltaic Efficiency on DSSC. *Mater. Chem. Phys.* **2010**, *122*, 284–289. [CrossRef]
13. Fang, X.; Li, M.; Guo, K.; Li, J.; Pan, M.; Bai, L.; Luoshan, M.; Zhao, X. Graphene Quantum Dots Optimization of Dye-Sensitized Solar Cells. *Electrochim. Acta* **2014**, *137*, 634–638. [CrossRef]
14. Gong, J.; Sumathy, K.; Qiao, Q.; Zhou, Z. Review on Dye-Sensitized Solar Cells (DSSCs): Advanced Techniques and Research Trends. *Renew. Sustain. Energy Rev.* **2017**, *68*, 234–246. [CrossRef]
15. Wang, L.; Jia, L.; Li, Q. A Novel Sulfur Source for Biosynthesis of (Ag, S)-Modified TiO_2 Photoanodes in DSSC. *Mater. Lett.* **2014**, *123*, 83–86. [CrossRef]
16. Jeong, H.; Lee, Y.; Kim, Y.; Kang, M. Enhanced Photoelectric Efficiency by Surface Modification of TiO_2 Thin Film Using Various Acidic Species. *Korean J. Chem. Eng.* **2010**, *27*, 1462–1468. [CrossRef]
17. Jin, E.M.; Zhao, X.G.; Park, J.-Y.; Gu, H.-B. Enhancement of the Photoelectric Performance of Dye-Sensitized Solar Cells Using Ag-doped TiO_2 Nanofibers in a TiO_2 Film as Electrode. *Nanoscale Res. Lett.* **2012**, *7*, 97. [CrossRef]
18. Shakir, S.; Abd-ur-Rehman, H.M.; Yunus, K.; Iwamoto, M.; Periasamy, V. Fabrication of Un-Doped and Magnesium Doped TiO_2 Films by Aerosol Assisted Chemical Vapor Deposition for Dye Sensitized Solar Cells. *J. Alloy. Compd.* **2018**, *737*, 740–747. [CrossRef]
19. Supriyanto, A.; Furqoni, L.; Nurosyid, F.; Hidayat, J.; Suryana, R. Effect of Sintering Temperatures and Screen Printing Types on TiO_2 Layers in DSSC Applications. *AIP Conf. Proc.* **2016**, *1717*, 040001. [CrossRef]
20. Noor, S.; Sajjad, S.; Leghari, S.A.K.; Shaheen, S.; Iqbal, A. ZnO/TiO_2 Nanocomposite Photoanode as an Effective UV-vis Responsive Dye Sensitized Solar Cell. *Mater. Res. Express* **2018**, *5*, 095905. [CrossRef]
21. Hamadanian, M.; Safaei-Ghomi, J.; Hosseinpour, M.; Masoomi, R.; Jabbari, V. Uses of New Natural Dye Photosensitizers in Fabrication of High Potential Dye-Sensitized Solar Cells (DSSCs). *Mater. Sci. Semicond. Process.* **2014**, *27*, 733–739. [CrossRef]
22. Cho, J.S.; Hong, Y.J.; Kang, Y.C. Electrochemical Properties of Fiber-in-Tube-and Filled-Structured TiO_2 Nanofiber Anode Materials for Lithium-Ion Batteries. *Chem.-Eur. J.* **2015**, *21*, 11082–11087. [CrossRef]
23. Jin, E.M.; Park, J.-Y.; Hwang, K.-J.; Gu, H.-B.; Jeong, S.M. Biotemplated Hybrid TiO_2 Nanoparticle and TiO_2–SiO_2 Nanocomposites for Dye-Sensitized Solar Cells. *Mater. Lett.* **2014**, *131*, 190–193. [CrossRef]
24. Desai, U.V.; Xu, C.; Wu, J.; Gao, D. Hybrid TiO_2-SnO_2 Nanotube Arrays for Dye-Sensitized Solar Cells. *J. Phys. Chem. C* **2013**, *117*, 3232–3239. [CrossRef]
25. Luan, X.; Wang, Y. Ultrathin Exfoliated TiO_2 Nanosheets Modified with ZrO_2 for Dye-Sensitized Solar Cells. *J. Phys. Chem. C* **2014**, *118*, 18917–18923. [CrossRef]
26. Jin, E.M.; Park, J.-Y.; Gu, H.-B.; Jeong, S.M. Synthesis of SnO_2 Hollow Fiber Using Kapok Biotemplate for Application in Dye-Sensitized Solar Cells. *Mater. Lett.* **2015**, *159*, 321–324. [CrossRef]
27. Swathy, K.S.; Abraham, P.A.; Panicker, N.R.; Pramanik, N.C.; Jacob, K.S. Nanostructured Anatase Titania Spheres as Light Scattering Layer in Dye-Sensitized Solar Cells. *Procedia Technol.* **2016**, *24*, 767–773. [CrossRef]
28. Shabdan, Y.; Ronasi, A.; Coulibaly, P.; Moniruddin, M.; Nuraje, N. Engineered Core-Shell Nanofibers for Electron Transport Study in Dye-Sensitized Solar Cells. *AIP Adv.* **2017**, *7*, 065008. [CrossRef]
29. Wang, H.; Wang, B.; Yu, J.; Hu, Y.; Xia, C.; Zhang, J.; Liu, R. Significant Enhancement of Power Conversion Efficiency for Dye Sensitized Solar Cell Using 1D/3D Network Nanostructures as Photoanodes. *Sci. Rep.* **2015**, *5*, 9305. [CrossRef]

30. Weintraub, B.; Wei, Y.; Wang, Z.L. Optical Fiber/Nanowire Hybrid Structures for Efficient Three-Dimensional Dye-Sensitized Solar Cells. *Angew. Chem. Int. Ed.* **2009**, *121*, 9143–9147. [CrossRef]
31. Ding, Y.; Zhou, L.; Mo, L.; Jiang, L.; Hu, L.; Li, Z.; Chen, S.; Dai, S. TiO_2 Microspheres with Controllable Surface Area and Porosity for Enhanced Light Harvesting and Electrolyte Diffusion in Dye-Sensitized Solar Cells. *Adv. Funct. Mater.* **2015**, *25*, 5946–5953. [CrossRef]
32. Ruqaishy, M.A.; Marzouqi, F.A.; Qi, K.; Liu, S.-Y.; Karthikeyan, S.; Kim, Y.; Al-Kindy, S.M.Z.; Kuvarega, A.T.; Selvaraj, R. Template-Free Preparation of TiO_2 Microspheres for the Photocatalytic Degradation of Organic Dye. *Korean J. Chem. Eng.* **2018**, *35*, 2283–2289. [CrossRef]
33. Choi, J.; Yoo, K.S.; Kim, J. Spray Pyrolysis Synthesis of Mesoporous TiO_2 Microspheres and Their Post Modification for Improved Photocatalytic Activity. *Korean J. Chem. Eng.* **2018**, *35*, 2480–2486. [CrossRef]
34. Park, N.-G.; Schlichthörl, G.; Van de Lagemaat, J.; Cheong, H.M.; Mascarenhas, A.; Frank, A.J. Dye-Sensitized TiO_2 Solar Cells: Structural and Photoelectrochemical Characterization of Nanocrystalline Electrodes Formed from the Hydrolysis of $TiCl_4$. *J. Phys. Chem. B* **1999**, *103*, 3308–3314. [CrossRef]
35. Liu, L.; Niu, H.; Zhang, S.Z.; Wan, L.; Miao, S.; Xu, J. Improved Performance of Dye-Sensitized Solar Cells: An TiO_2–Nano-SiO_2 Hybrid Photoanode with Post-Treatment of $TiCl_4$ Aqueous Solution. *Appl. Surf. Sci.* **2012**, *261*, 8–13. [CrossRef]

nanomaterials

MDPI

Article

Surface Composition and Preparation Method for Oxygen Nanobubbles for Drug Delivery and Ultrasound Imaging Applications

Muhammad Saad Khan [1,†], Jangsun Hwang [1,†], Kyungwoo Lee [1], Yonghyun Choi [1], Jaehee Jang [1], Yejin Kwon [1], Jong Wook Hong [2,3,*] and Jonghoon Choi [1,*]

1 School of Integrative Engineering, Chung-Ang University, Seoul 06974, Korea; saad.engr@gmail.com (M.S.K.); isnickawesome@gmail.com (J.H.); orztapa@gmail.com (K.L.); dydgus5057@gmail.com (Y.C.); jjaeh95@gmail.com (J.J.); angang1027@gmail.com (Y.K.)

2 Department of Bionano Technology, Hanyang University, Seoul, Korea

3 Department of Bionano Engineering, Hanyang University, Ansan, Korea

* Correspondence: jwh@hanyang.ac.kr (J.W.H.); nanomed@cau.ac.kr (J.C.); Tel.: +82-31-400-5206 (J.W.H.); +82-2-820-5258 (J.C.)

† These authors contributed equally to this work.

Received: 3 December 2018; Accepted: 23 December 2018; Published: 2 January 2019

Abstract: Phospholipids have been widely investigated for the preparation of liposomes, and micro and nanobubbles. They comprise biocompatible and biodegradable molecules and offer simple preparation with a variety of functions in diagnostic and therapeutic applications. Phospholipids require emulsifiers and surfactants to assemble in the form of bubbles. These surfactants determine the size, zeta potential, and other characteristics of particles. Polyethylene glycol (PEG) and its various derivatives have been employed by researchers to synthesize micro and nanobubbles. The stability of phospholipid-shelled nanobubbles has been reported by various researchers owing to the reduction of surface tension by surfactants in the shell. Nanobubbles have been employed to deliver oxygen to tissues and hypoxic cells. In this study, we investigated the effects of different ratios of phospholipids to PEG on the size, distribution, and characterization of oxygen nanobubbles (ONBs). ONBs were synthesized using a sonication technique. We analyzed and compared the sizes, numbers of generated particles, and zeta potentials of different compositions of ONBs using dynamic light scattering and nanoparticle tracking analysis. Then, we employed these oxygen nanobubbles to enhance the cellular microenvironment and cell viability. ONBs were also investigated for ultrasound imaging.

Keywords: oxygen nanobubbles; phospholipids; polyethylene glycol; ultrasound imaging

1. Introduction

Novel bionanotechnology employing engineered micro and nano structures has been developed for diagnostic and therapeutic applications in the field of healthcare [1]. Micro and nanobubbles are one of those new assemblies already in use in a variety of medical applications, typically as ultrasound contrast agents. Microbubbles are inherently echogenic, owing to a mismatch of acoustic impedance at the liquid/gas interface, which provides an ultrasonic contrast [2,3]. Although large bubbles would generate a greater contrast in ultrasonic imaging, the size of blood capillaries limits the microbubble size to less than 10 μm. Therefore, microbubbles in the range of 1 to 8 μm are preferred for medical usage in ultrasound contrast agents because they have a resonance frequency in the range of 1 to 12 MHz, which is usually employed in commercial applications [2,4–9].

Recently, nanobubbles have been investigated for theranostic (or theragnostic) applications, which involve combinations of diagnostic and therapeutic applications. Nanobubbles mostly consist of similar shell/core compositions, in which the shell is often formed of a unilaminar composition of phospholipids, polymers, or proteins, and the core contains a less soluble gas. In contrast to bilayer liposomes of a similar size, nanobubbles have a monolayer shell with a less-soluble gas core, and, therefore, they can be utilized for gas delivery applications [8,10–12]. Nanobubbles are of particular interest, owing to their smaller size, biodegradability, higher surface area, longer half-life, higher cellular uptake, and echogenic properties [13]. Owing to their size in the nanometer range, nanobubbles have been investigated for passive targeting in normal tissues and tumors, considering the opportunity to exploit the enhanced permeability and retention effect [5,14–18]. Yin et al. demonstrated that nanobubbles exhibited similar echogenic properties to microbubbles when high ultrasound frequencies were employed, and nanobubbles were retained in tumors for longer periods compared with microbubbles [19].

The stability of the bubbles is governed by the Laplace pressure, represented in the following equation [2,13,20,21]:

$\Delta P = P_{in} - P_{out} = 2\frac{\delta}{r}$ where δ is the interfacial tension, and r is the radius of the bubble. The Laplace pressure depends on the size of the bubble and the interfacial pressure at the surface. As indicated by the equation, smaller bubbles experience a higher pressure difference, which leads to their rupture. This equation has been discussed for its application in nanosized bubbles by various researchers [13,22–24]. Cavalli et al. [13] discussed the role of surfactants in rthe eduction of the value of δ to stabilize the bubbles at nanosize. Less soluble gas in the core can also promote the stability of the bubbles. Experimental evidences and characterizations by various researchers reveal that nanobubbles can be used in diagnostic and applications [12,16,17,19,25,26]. The main role in the stability of the nanobubbles is the composition of shell and surfactants. Surfactants reduce the external Laplace pressure, and, therefore, nano-sized nanobubbles can remain stable if an appropriate ratio of surfactants are in the shell. It is important to investigate the optimal ratio of surfactants and their role in the size and number of bubbles [2,27,28].

Most drug and gas delivery applications require the key characteristics of biocompatibility, biodegradability, ease of transport of molecules, cellular attachment, and drug loading abilities [29]. The composition of the shell plays an important role in determining the stiffness, echogenic response, permeability to diffusion, and renal clearance [2]. Phospholipids have an amphiphilic character, and they contain hydrophobic chains and hydrophilic headgroups, which give them a unique property of self-assembly around hydrophobic gas in the core in the presence of surfactants and emulsifiers. Owing to this self-assembly characteristic, phospholipid-shelled bubbles can be synthesized using a variety of techniques, including sonication, agitation, microfluidics, and emulsification. Acyl chains provide cohesiveness and stability to the phospholipids. The longer the acyl chains, the higher the stability of the bubbles. Various compositions have been employed by the researchers for synthesizing micro- and nanobubbles. The phase transition temperature is an important parameter for phospholipids because at this temperature the phospholipids undergo a transition from gel to liquid and their fluidity increases. The phase transition temperature also depends on the polar head group, the degree of saturation of hydrocarbon chains, and the acyl chain lengths. Nanobubbles are prepared above phase transition temperatures to achieve a homogenous composition, while they are stored below phase transition temperatures after preparation [7,30,31]. Phospholipids are biocompatible molecules, and various researchers have reported their non-toxicity. Nano-sized phospholipids tend to be more stable. Various lipid formulations have been reported by researchers for gas and drug delivery applications [3,32–35].

Polyethylene glycol (PEG) is a flexible, hydrophilic polymer, and it has been employed in various research studies as an emulsifier in the preparation of micro and nanobubbles and other biological applications [36–42]. PEGylation contributes considerably to imparting biocompatibility and enhancing the lifespan of nanobubbles. PEG has been incorporated into various compositions of lipids, polymers, and proteins to synthesize micro/nanobubbles and liposomes [39,40,43]. PEG derivatives can be utilized to incorporate functional groups into the shells of nanobubbles. The ratio of PEG components is critical for the preparation of nanobubbles, as it determines the sizes and characteristics of these particles.

The zeta potential plays an important role in the preparation of nanobubbles and liposomes as it determines the stability of particles, their cellular intake, and their resistance against coalescence. A cell membrane has a negative potential, and logically it might attract higher positively charged particles. However, higher negative particles with a zeta potential in the range of −45 mV have also been reported to exhibit a greater cellular uptake [31]. On the other hand, gene delivery applications utilize cationic lipids with a positive charge to form complexes with negatively charged genes [21,44].

Although ultrasound-mediated gas and drug delivery are gaining more attention, nanobubbles can be employed without employing ultrasound. Phospholipid shells are permeable to gas, and as the gas diffuses out as the bubble shrinks and eventually breaks owing to the increasing external Laplace pressure, thereby releasing the encapsulated drug [7].

Oxygen delivery through micro and nanobubbles is a key concept aimed at the reversal of hypoxia in tumor microenvironment gas and other drug delivery applications [27,28,34,45–48]. Nanobubbles can be utilized to create supersaturated fluids for oxygen delivery. In a separate study, we demonstrated the use of oxygen nanobubbles for the effective reversal of hypoxia [8]. We demonstrated that oxygen nanobubbles (ONBs) can be used for downregulation of the hypoxia-inducible factor-1α (HIF-1α) protein, which is of clinical significance in chemotherapy. Higher oxygen availability also favors the therapeutic efficacy of radiotherapy and photodynamic therapy [7] and, therefore, ONBs have many potential therapeutic applications.

This study aims to identify the role of the composition and surfactants in the preparation of nanobubbles. To study this, we initially experimented with three different compositions to investigate the size, number of produced particles, and zeta potential of the nanobubbles. In further experiments, we selected one composition ratio and continued to evaluate its characterization, cytotoxicity, and ultrasound imaging. We were keen to investigate the role of the PEG surfactant on the size and zeta potential of oxygen nanobubbles. We found that ratios in the range of 90:10 and 80:20 are optimal, and, thus, we finally opted for 85:15.

2. Materials and Methods

2.1. Materials

The phospholipids 1,2-distearoyl-*sn*-glycero-3-phosphocholine (DSPC), 1,2-distearoyl-*sn*-glycero-3-phosphoethanolamine-*N*-[amino(polyethylene glycol)-2000] (ammonium salt) (DSPE-PEG-2000-Amine), and 1,2-distearoyl-*sn*-glycero-3-phosphoethanolamine-*N*-[biotinyl(polyethylene glycol)-2000] (ammonium salt) (DSPE-PEG-2000-Biotin) were purchased from Avanti Polar Lipids (Merck, Kenilworth, NJ, USA), while Fluorescein-avidin conjugate (FITC-avidin) was purchased from Thermo Fisher Scientific (Waltham, MA, USA). Dulbecco's phosphate-buffered saline (DPBS) and chloroform were purchased from Sigma Aldrich (c St. Louis, MO, USA).

2.2. Preparation of Oxygen Nanobubbles

DSPC, DSPE-PEG-2000-Amine, and DSPE-PEG-2000-Biotin were dissolved in chloroform in molar ratios of 50:50:0, 80:20:0, 85:8:7, and 90:5:5. The chloroform was dried by heating the flask in a hot air oven at 80 °C, which resulted in a thin dried lipid layer using 10 mL DPBS. The total final concentration of lipids in all composition ratios was approximately 4.1 mg/mL. To dissolve these lipids in DPBS, a bath tub sonicator was heated to 60 °C, and sonication was performed until the entire lipid layer was dispersed in DPBS, forming a milky suspension. Next, the suspension was further sonicated with a tip-sonicator at 190 W for 5 min in the presence of a 99.9% oxygen supply via a cylinder to synthesize ONBs. Fluorescent bubbles were prepared by adding 100 μL of FITC-avidin to 10 mL suspension and centrifuging at 500 g for 10 min.

2.3. Characterization of Oxygen Nanobubbles

For the purpose of imaging, bright-field and fluorescence microscopy, transmission electron microscopy (TEM), and scanning electron microscopy (SEM) were employed. Micro-sized bubbles were collected for microscopic imaging, while SEM was performed for both micro- and nano-sized bubbles. TEM was employed for nano-size imaging.

Samples for SEM were created by injecting microbubbles in 5% agarose gel. First, the agarose gel was heated to 100 °C. Then, ONBs were added during the cooling process when the temperature of the agarose gel was below 50 °C. The agarose gel was allowed to dry, and then thin layers of the dried gel were cut for imaging purposes. These layers were coated with Pt for SEM imaging.

Samples for TEM were prepared by negative staining using uranyl acetate. First, a copper grid was dipped in nanobubbles and dried. Then, it was washed and negatively stained with 2% uranyl acetate solution. These grids were then utilized for TEM imaging.

Dynamic light scattering (DLS) (Malvern, PA, USA) was also employed to analyze the size and zeta potential of nanobubbles. Samples were diluted at 1:100 for DLS reading. The zeta potential was measured using a disposable capillary cell (DTS1070). Nanoparticle tracking analysis (NTA) (Nanosight NS300, Malvern, PA, USA) was employed for counting number of particles/mL.

2.4. Cytotoxicity Tests

Cytotoxicity tests were conducted using NIH-3T3, MDA-MB-231 breast cancer cells, Chinese hamster ovary (CHO) cells, and human adipose derived stem cells (HADSCs). 1×10^4 cells per well were seeded in a 96-well plate, and CCK-8 was used to determine cell viability after treatment was carried out with various concentrations of nanobubbles and lipid solution over a period of 24 h.

For a hemolysis test, different concentrations of ONBs (0.1, 5, 10, 100, and 200 μL/mL) and respective lipid constituents were added to 1 mL of sheep blood. The mixture was kept in an incubator for 24 h. Drakin's solution was added in a 96-well plate, and the measurements were performed at 510 nm through a microplate reader. 10% Tryton-X100 was employed as a positive control.

2.5. Ultrasound Imaging

ONBs were inserted (dilution 1:10 in DPBS) in a balloon and compared with a tap water sample using a commercially available general electric (GE) ultrasound machine for three different transducers with frequencies of 3–5 MHz, 6–9 MHz and 6–15 MHz, respectively.

2.6. Statistical Analysis

GraphPad Prism software was used for statistical analysis and graphical representations of the data. T-tests were performed on the data, and the significance was checked. Non-significant values are indicated by ns in the results section, while the symbols "*, **, ***, and **** " indicate *p*-values of less than 0.05, 0.01, 0.001, and 0.0001, respectively.

3. Results

Table 1 presents the compositions of samples in terms of the mass and molar ratio used during the experiments. We used DSPE-PEG-Biotin in two combinations to test the avidin/biotin interaction in the case of fluorescence ONBs. The total final concentration of lipids was kept at approximately 4.1 mg/mL by rehydrating the dried lipids in 10 mL DPBS.

Figure 1 illustrates the impact of the composition ratio on the size, the number of particles, zeta potential, and size distribution for the three composition ratios of 90:5:5, 80:20:0, and 50:50:0. Figure 1A shows that the number of particles decreases as we decrease the base lipid DSPC and increase the PEG component of the ONBs. In the case of 90:5:5, the number of particles is 4.77×10^{11} particles/mL, while in case of 80:20:0 it is 3.4×10^{11} particles/mL, which is 27% less.

For the sample with a 50:50:0 composition, the number of particles is 1.88×10^{11} /mL, which is approximately 60% less than for the 90:5:5 sample. Figure 1B illustrates the impact of the composition ratio on the size. The ratio 90:5:5 exhibits a mean size of 394 ± 79 nm, 80:20:0 has a mean size of 246 ± 46 nm, and 50:50:0 has a mean size of 173 ± 25 nm. This reduction in size can be attributed to the high surfactant ratio.

Table 1. Composition ratios of phospholipid and surfactants.

Sample #	Combination	Mass (mg)	Molar Ratio
	DSPC	29.53	90
1	DSPE-PEG-2000-Amine	5.8	5
	DSPE-PEG-2000-Biotin	6.26	5
	DSPC	22.02	80
2	DSPE-PEG-2000-Amine	19.5	20
	DSPE-PEG-2000-Biotin	0	0
	DSPC	9.08	50
3	DSPE-PEG-2000-Amine	32.39	50
	DSPE-PEG-2000-Biotin	0	0
	DSPC	25.2	85
4	DSPE-PEG-2000-Amine	8.4	8
	DSPE-PEG-2000-Biotin	7.9	7

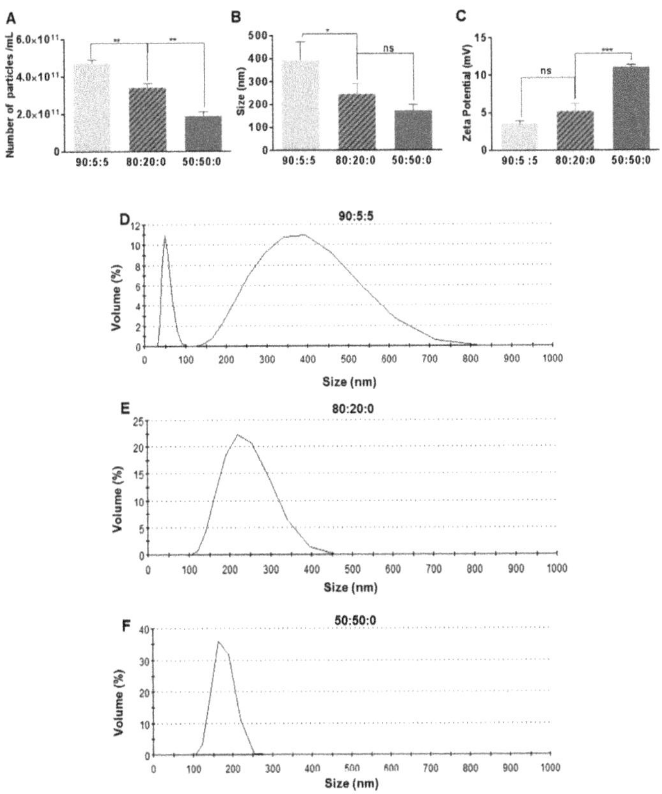

Figure 1. Impact of various compositions on the number (**A**), the size (**B**) and the zeta potential (**C**) of nanobubbles. Dynamic light scattering (DLS) data on the size distribution for 90:5:5 (**D**), for 80:20:0 (**E**), and for 50:50:0 (**F**). Here, *, **, ***, andindicate p-values of less than 0.05, 0.01, 0.001, and 0.0001, respectively.

Figure 1C shows the increase in the zeta potential owing to an increase in the PEG ratio. The PEG surfactant contained an amine derivative, which might be responsible for the higher positive charge in the cases of 80:20:0 and 50:50:0. Figure 1D–F show the DLS data for the size distribution for the 90:5:5, 80:20:0, and 50:50:0 cases, respectively. The size distribution range for the 90:5:5 composition ratio exhibits two peaks: One at approximately 50 nm and another broad distribution from 150 nm to 750 nm with the peak containing approximately 10% of particles (Figure 1D). In the case of 80:20:0, the size range narrows to 100–400 nm and the peak contains approximately 25% of particles (Figure 1E). In the case of 50:50:0, the size range is from 100 nm to 250 nm (Figure 1F), which is significantly narrower than the two previous compositions, and the peak contains approximately 35% of particles, which is higher than the other two compositions. It can be observed that a higher PEG content leads to a narrower size distribution and decreased average size, as also indicated in Figure 1B.

Figure 2 illustrates the characterization of ONBs in the 85:8:7 composition ratio. Figure 2A shows a TEM image of ONBs. Figure 2B depicts an SEM image. These images contain both micro-sized and nano-sized particles. Figure 2C shows MDA-MB-231 cells injected with ONBs, captured through an optical microscope. Figure 2D shows an image of micro-sized particles on a glass slide in optical microscopy, while Figure 2E presents a fluorescent image of ONBs when FITC-avidin was conjugated with ONBs through an avidin-biotin interaction. Figure 2F depicts the size distribution of ONBs obtained through DLS. The mean size is approximately 300 nm, while the distribution range is mainly between 200 and 500 nm.

Figure 3 illustrates the cell viability against various concentrations of ONBs and lipid constituents in various cell lines. Figure 3A shows the cell viability of ONBs and lipids in NIH-3T3 cells. The control ONB group has a cell viability of 94%, while 1 μL/mL and 10 μL/mL exhibit similar cell viabilities. However, in the 50 μL/mL case, cell viability is enhanced by 15% compared with that in the control case, and for 100 μL/mL, cell viability is enhanced by 7%. Conversely, the lipid constituents reduced cell viability by approximately 20% when they were utilized in 50 μL/mL and 100 μL/mL concentrations. This result indicates that a higher concentration of lipids reduces cell viability in NIH-3T3. However, ONBs in concentrations of 50 μL/mL and 100 μL/mL improved the cell viability by more than 30% as compared to lipids of similar concentrations, which can be attributed to the presence of oxygen and higher availability of oxygen.

Figure 3B illustrates the cell viability in the case of MDA-MB-231 breast cancer cells. This indicates that the 1 μL/mL and 10 μL/mL concentrations exhibit similar cell viability as compared to the control group, while a concentration of 50 μL/mL enhances cell viability by 5%. In the case of MDA-MB-231, the lipid constituents do not exhibit any cytotoxicity, even at higher concentrations of 50 μL/mL and 100 μL/mL. When we compare this result with other cell lines, it is evident that tumor cells exhibit more tolerance to higher lipid concentrations. While 50 μL/mL ONBs exhibited an increase in cell viability, 100 μL/mL ONBs did not exhibit any significant difference during 24 h of the experiment. Therefore, 50 μL/mL proved to be more effective in this particular case.

Figure 3C shows the cell viability assay of CHO cells. In this case, 1 μL/mL and 10 μL/mL ONBs increased cell viability by approximately 20% compared with the ONB control, while concentrations of c increased cell viability by approximately 15%. The lipid constituents in this group did not cause a significant increase or decrease in cell viability compared with that in the control group. Higher concentrations of lipids, i.e., 50 μL/mL and 100 μL/mL, also did not exhibit any significant toxicity.

Figure 3D shows the cell viability for HADSCs. Concentrations of 1 μL/mL and 10 μL/mL ONBs enhanced cell viability, while a concentration of 50 μL/mL reduced cell viability. In the case of lipids, 50 μL/mL and 100 μL/mL significantly reduced cell viability. The reason can be attributed to the lipid concentration.

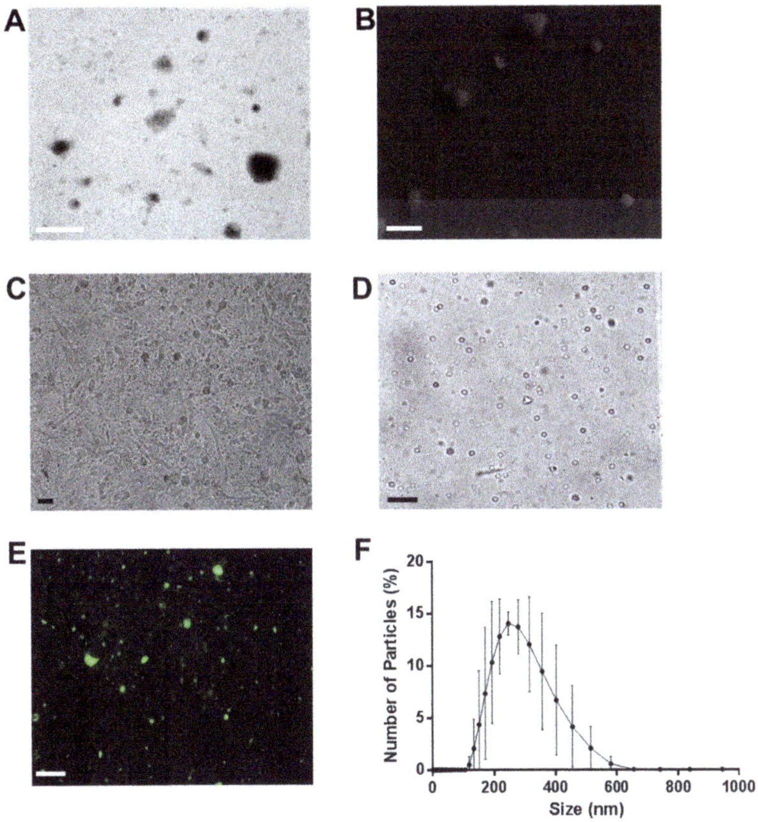

Figure 2. Characterization of oxygen nanobubbles (ONBs) with a lipid ratio of 85:8:7. (**A**) Transmission electron microscopy (TEM) image, with a scale bar of 500 nm. (**B**) Scanning electron microscopy (SEM) image, with a scale bar of 1 μm. (**C**) MDA-MB-231 cells with 10% ONBs in the media and a scale bar of 10 μm. (**D**) Micro-sized bubbles with a scale bar of 10 μm. (**E**) Fluorescent bubbles with a scale bar of 10 μm. (**F**) Combined DLS data of ONBs with $n = 6$.

Figure 3E shows the result of the hemolysis experiment. The ONBs exhibited less than 2% hemolysis in all cases between 1 μL/mL and 200 μL/mL. In the case of lipid constituents, similar results were observed. From these results, we can observe that ONBs do not cause hemolysis.

Figure 4 illustrates the ultrasonic imaged of ONBs. Figure 4A show ultrasonic images captured with a 3 to 5 MHz curved transducer. We captured the screens with no sample, tap water, and nanobubbles, respectively. There is no visible difference between the tap water and nanobubbles. Since most micro size bubbles have a resonance frequency in the range of 3 to 5 MHz, these results suggest that our sample did not include many microsize particles to exhibit higher contrast as compared to tap water. It also indicates that nanobubbles do not generate significant echoes compared with tap water. Figure 4B show images acquired with a 6 to 9 MHz linear probe. It can be observed that the contrast provided by nanobubbles is slightly improved compared with tap water. Figure 4C present images captured with a 6 to 15 MHz transducer. The difference between the nanobubbles and tap water is visible. Thus, we can confirm that nanobubbles generate echoes at higher frequencies, and they can be traced using commercial ultrasound equipment. The mechanical index (MI) during these images was kept constant at 0.15. White circles in the figures indicate the area of a balloon containing ONBs.

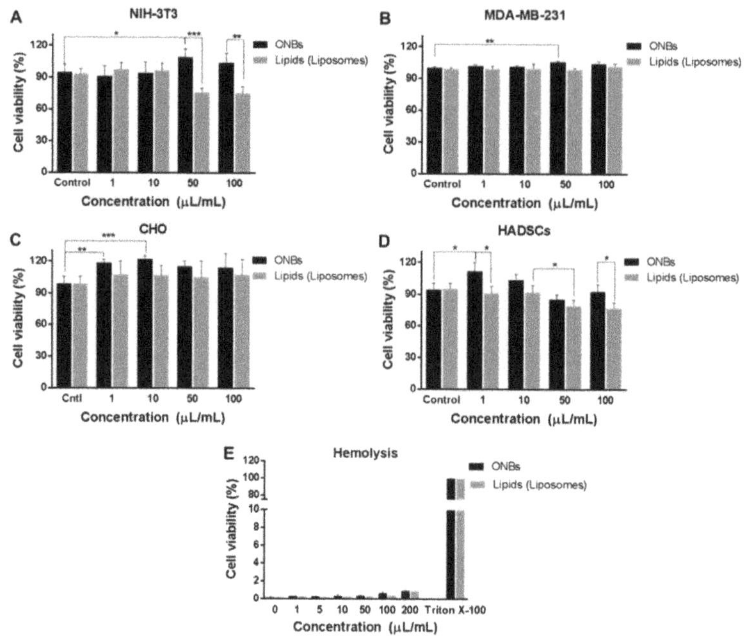

Figure 3. Cell viability test for ONBs determined through CCK-8. Cell viability of various concentrations of ONBs and lipids in NIH-3T3 (**A**) in MDA-MB-231 (**B**), in CHO (**C**) and in HADSCs (**D**) cell lines. (**E**) Hemolysis experiment for ONBs and lipids. Here, *, **, and *** highlight *p*-values of less than 0.05, 0.01, and 0.001, respectively.

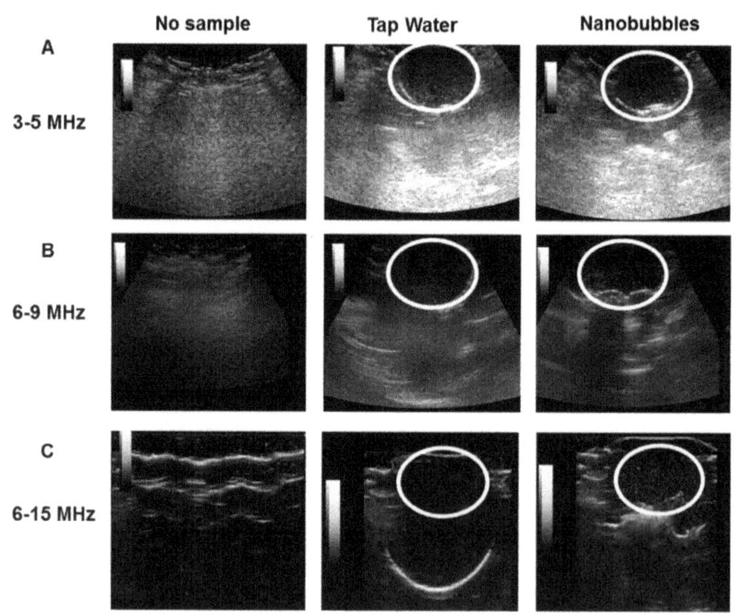

Figure 4. Ultrasound images of ONBs. No sample, tap water, and nanobubbles were compared. (**A**) shows images with a 3 to 5 MHz transducer. (**B**) shows images for a 6 to 9 MHz transducer. (**C**) shows images with a 6 to 15 MHz transducer.

4. Discussion

In this study, we focused on investigating the effects of the shell composition on the size and characteristics of ONBs. We found that increasing the PEG ratio reduced the sizes of the bubbles and narrowed their distribution. The distribution range in the case of 90:5:5 was between 50 and 750 nm; this narrowed to 100 to 400 nm in the case of 80:20:0 and 100 to 250 nm in the case of 50:50:0. Reducing the size and narrowing down the size distribution is a desirable effect in nanobubble preparation because that way the enhanced permeability and retention (EPR) effect might be utilized for higher cellular uptake in theranostic applications. However, a higher PEG ratio also reduced the number of synthesized particles/mL during the process. The number of particles in the cases of 80:20:0 and 50:50:0 were reduced by 27% and 60%, respectively, compared with the 90:5:5 composition. Therefore, it can be concluded that a higher ratio of base phospholipid (>80%) is optimal for generating the higher number of particles. The zeta potential also considerably increased in the case of a 50:50:0 composition ratio compared with those of 90:5:5 and 80:20:0. Although higher zeta potential is favorable for some applications like gene delivery, other applications might prefer closer to neutral potential.

To optimize the composition ratio to obtain an optimal number of particles and narrow size distribution, we selected an 85:15 molar ratio containing 85% DSPC and 15% surfactant. The reason for this selection is that it imparts more functionality options by containing both amine and biotin functional groups, and we were able to synthesize fluorescent ONBs. By maintaining this composition ratio, we were also able to generate a higher number of particles in a narrower distribution. The 85:8:7 composition resulted in nanobubbles with a peak size of approximately 300 nm and distribution range of 200 to 400 nm, while the number of particles was 4.11×10^{11} particles/mL. This represents a tradeoff between a smaller size range and the number of generated bubbles. TEM and SEM imaging were conducted to observe the characteristics of ONBs. TEM image showed several ONBs in the size range of 100 nm or smaller. That might be due to the vacuuming effect during imaging. SEM image indicated the surface morphology, some of the particles were not spherical which might be the result of aggregation during sample preparation. Therefore, microsize particles were collected for optical microscopy and the bubbles are clearly observable with shell/core composition. The microscopic image of ONBs with MDA-MB-231 cells also indicate many ONBs in the cellular media. Biotin functional group was evaluated by synthesizing fluorescent ONBs using FITC-avidin.

Phospholipids are generally biocompatible, and PEG derivatives enhance the biocompatibility. PEGylation improves the stability and half-life of nanobubbles. It also imparts a resistance to coalescence. We tested the cell viability of ONBs in different cell lines, and it can be concluded that they are not cytotoxic for cells. The total concentration of phospholipids and PEG derivatives were kept approximately 4.1 mg/mL for all experiments and we compared the biocompatibility of ONBs with different lipid constituents (DSPC, DSPE-PEG-2000-amine, and DSPE-PEG-2000-biotin) to determine the role of oxygen in the biocompatibility and cytotoxicity. It is clearly observable that in most cases ONBs either improved or maintained cell viability. Even at higher concentrations, the cell viability in the case of ONBs was higher compared with the lipid counterparts. This clearly indicates that oxygen plays a positive role in enhancing cell viability. As indicated in the results for the cases of NIH-3T3 and AHDSCs, 50 μL/mL (5% *v*/*v*) and 100 μL/mL (10% *v*/*v*) ONBs exhibited higher cell viabilities compared with lipids of the same concentrations. Liposomes in these conditions reduced cell viability at 5% and 10% concentrations, which can be attributed to access amount of lipids damaging the cells.

In the case of the breast cancer cell line MDA-MB-231, ONBs and lipids did not exhibit any cytotoxicity, while 5% ONBs exhibited an increase in cell viability as compared to the control group. This increase in cell viability may be attributed to a higher level of oxygen present which enhances cell differentiation. While 100 μL/mL ONBs did not increase cell viability, but they did not exhibit any cytotoxicity either in this case. A higher level of oxygen has been associated with downregulation of HIF-1α protein in tumors which is critical for tumor treatment. Further studies are required to investigate the optimal level of oxygen for tumor cells to enhance treatment. Tumor cells are an important target of drug delivery applications using nanobubbles, and our results exhibit that higher concentrations of ONBs and liposomes up to 10% of media volume can be employed in such cases for drug delivery. This is an important aspect as more drug can be loaded into the higher number of particles for anti-tumor drug delivery applications.

In the case of CHO cell line, 1 µL /mL (0.1% *v*/*v*) and 10 µL/mL (1% *v*/*v*) ONBs were more beneficial for the enhancement of cell viability. Higher concentrations of ONBs and lipids, i.e., 50 µL/mL and 100 µL/mL did not cause a significant reduction in cell viability. All these results demonstrate that ONBs can be employed for delivering oxygen to cells and that they improve cell viability by enhancing the microenvironments of cells.

Directly exposing blood to oxygen pressure may cause hemolysis, but ONBs do not cause hemolysis even at 20% (*v*/*v*), as shown by our results. This result is significant because hemolysis is a major issue when blood is directly exposed to oxygen pressure. In our study, ONBs released oxygen through a diffusion mechanism, which is a slow natural process, and, therefore, hemolysis did not occur. Therefore, it can be concluded that a higher level of oxygen (ONBs 10% of media volume) is generally not harmful to the cellular microenvironment and instead, it helps in improving the cell viability in most cases. Further, ONBs can be used to supply oxygen more than tissue partial pressure, and this method can be effectively used for reversal of hypoxic conditions in tumors as demonstrated previously by our group and other researchers. This aspect is of significant importance in anti-tumor applications related to chemotherapy, radiotherapy, and photodynamic therapy.

Ultrasonic imaging with commercial ultrasound machine revealed that these ONBs can be traced using ultrasound. Nanobubbles exhibited a higher contrast compared with tap water in ultrasound imaging at higher frequencies (6–15 MHz transducer), but this was not a very clear contrast difference, as is required for diagnostic applications. However, these bubbles can be traced using ultrasound, and, therefore, ultrasound can be employed for therapeutic purposes. At lower frequencies, in the range of 3 to 5 MHz, the contrast was not very significant compared with tap water. This is a further indication that our sample consisted of nano-sized particles and was generating a contrast at higher frequencies of ultrasound. This is more meaningful when nanobubbles are used for theranostic applications in combination with ultrasound.

5. Conclusions

Nanobubbles are a promising tool for diagnostic and therapeutic applications, and considerable research attention has been given to various applications of nanobubbles in medical fields. In this study, we summarized the role of the composition on the size and characteristics of oxygen nanobubbles, by employing different ratios of phospholipids and emulsifying PEG-derivatives. Increasing the PEG ratio was found to contribute to a reduction in size and distribution. However, a higher ratio of base lipids was required to generate a higher number of particles. The role of ONBs in the enhancement of cell viability was evaluated for various cell lines. Oxygen was supplied through ONBs, which were found to be non-toxic, and did not result in hemolysis in sheep blood. ONBs were employed in ultrasonic imaging, and it was observed that these ONBs can be traced using commercially available ultrasound transducers. Future studies are required to evaluate the impact of oxygen nanobubbles in tumor environment for longer time duration.

Author Contributions: M.S.K., J.H., J.W.H. and J.C. designed the experiments. M.S.K., J.H., K.L., Y.C. performed the experiments. M.S.K., J.H., J.J., Y.K. and J.C. analyzed the data. M.S.K., J.W.H. and J.C. wrote the manuscript.

Funding: This study was supported by the Nano Material Technology Development Program through the National Research Foundation of Korea (NRF) funded by the Ministry of Science and ICT (No. 2017M3A7B8061942). This research was also supported by the Basic Science Research Program 2018R1A2B6005354 through the National Research Foundation of Korea (NRF) funded by the Ministry of Science, ICT and Future Planning of Korea.

Conflicts of Interest: The authors declare no conflict of interest.

References

1. Choi, J.; Lee, E.K.; Choo, J.; Yuh, J.; Hong, J.W. Micro 3d cell culture systems for cellular behavior studies: Culture matrices, devices, substrates, and in-situ sensing methods. *Biotechnol. J.* **2015**, *10*, 1682–1688. [CrossRef]
2. Cavalli, R.; Bisazza, A.; Lembo, D. Micro-and nanobubbles: A versatile non-viral platform for gene delivery. *Int. J. Pharm.* **2013**, *456*, 437–445. [CrossRef]
3. Sirsi, S.; Borden, M. Microbubble compositions, properties and biomedical applications. *Bubble Sci. Eng. Technol.* **2009**, *1*, 3–17. [CrossRef]
4. Dijkmans, P.A.; Juffermans, L.J.; Musters, R.J.; van Wamel, A.; ten Cate, F.J.; van Gilst, W.; Visser, C.A.; de Jong, N.; Kamp, O. Microbubbles and ultrasound: From diagnosis to therapy. *Eur. J. Echocardiogr.* **2004**, *5*, 245–256. [CrossRef]
5. Fix, S.M.; Borden, M.A.; Dayton, P.A. Therapeutic gas delivery via microbubbles and liposomes. *J. Controll. Release* **2015**, *209*, 139–149. [CrossRef]
6. Cavalieri, F.; Finelli, I.; Tortora, M.; Mozetic, P.; Chiessi, E.; Polizio, F.; Brismar, T.B.; Paradossi, G. Polymer microbubbles as diagnostic and therapeutic gas delivery device. *Chem. Mater.* **2008**, *20*, 3254–3258. [CrossRef]
7. Khan, M.; Hwang, J.; Lee, K.; Choi, Y.; Kim, K.; Koo, H.-J.; Hong, J.; Choi, J. Oxygen-carrying micro/nanobubbles: Composition, synthesis techniques and potential prospects in photo-triggered theranostics. *Molecules* **2018**, *23*, 2210. [CrossRef]
8. Khan, M.S.; Hwang, J.; Seo, Y.; Shin, K.; Lee, K.; Park, C.; Choi, Y.; Hong, J.W.; Choi, J. Engineering oxygen nanobubbles for the effective reversal of hypoxia. *Artif. Cells Nanomed. Biotechnol.* **2018**, 1–10. [CrossRef]
9. Qin, S.; Caskey, C.F.; Ferrara, K.W. Ultrasound contrast microbubbles in imaging and therapy: Physical principles and engineering. *Phys. Med. Biol.* **2009**, *54*, R27. [CrossRef]
10. Pitt, W.G.; Husseini, G.A.; Staples, B.J. Ultrasonic drug delivery—A general review. *Expert Opin. Drug Deliv.* **2004**, *1*, 37–56. [CrossRef]
11. Koshiyama, K.; Wada, S. Collapse of a lipid-coated nanobubble and subsequent liposome formation. *Sci. Rep.* **2016**, *6*, 28164. [CrossRef]
12. Bhandari, P.; Wang, X.; Irudayaraj, J. Oxygen nanobubble tracking by light scattering in single cells and tissues. *ACS Nano* **2017**, *11*, 2682–2688. [CrossRef]
13. Cavalli, R.; Soster, M.; Argenziano, M. Nanobubbles: A promising efficienft tool for therapeutic delivery. *Ther. Deliv.* **2016**, *7*, 117–138. [CrossRef]
14. Xing, Z.; Wang, J.; Ke, H.; Zhao, B.; Yue, X.; Dai, Z.; Liu, J. The fabrication of novel nanobubble ultrasound contrast agent for potential tumor imaging. *Nanotechnology* **2010**, *21*, 145607. [CrossRef]
15. Wu, H.; Rognin, N.G.; Krupka, T.M.; Solorio, L.; Yoshiara, H.; Guenette, G.; Sanders, C.; Kamiyama, N.; Exner, A.A. Acoustic characterization and pharmacokinetic analyses of new nanobubble ultrasound contrast agents. *Ultrasound Med. Biol.* **2013**, *39*, 2137–2146. [CrossRef]
16. Zhou, M.; Cavalieri, F.; Caruso, F.; Ashokkumar, M. Confinement of acoustic cavitation for the synthesis of protein-shelled nanobubbles for diagnostics and nucleic acid delivery. *ACS Macro Lett.* **2012**, *1*, 853–856. [CrossRef]
17. Wang, Y.; Li, X.; Zhou, Y.; Huang, P.; Xu, Y. Preparation of nanobubbles for ultrasound imaging and intracelluar drug delivery. *Int. J. Pharm.* **2010**, *384*, 148–153. [CrossRef]
18. Choi, J.; Tung, S.-H.; Wang, N.S.; Reipa, V. Small-angle neutron scattering measurement of silicon nanoparticle size. *Nanotechnology* **2008**, *19*, 085715. [CrossRef]
19. Yin, T.; Wang, P.; Zheng, R.; Zheng, B.; Cheng, D.; Zhang, X.; Shuai, X. Nanobubbles for enhanced ultrasound imaging of tumors. *Int. J. Nanomed.* **2012**, *7*, 895.
20. Lee, M.; Lee, E.Y.; Lee, D.; Park, B.J. Stabilization and fabrication of microbubbles: Applications for medical purposes and functional materials. *Soft Matter* **2015**, *11*, 2067–2079. [CrossRef]
21. Ferrara, K.; Pollard, R.; Borden, M. Ultrasound microbubble contrast agents: Fundamentals and application to gene and drug delivery. *Annu. Rev. Biomed. Eng.* **2007**, *9*, 415–447. [CrossRef]
22. Attard, P.; Moody, M.P.; Tyrrell, J.W. Nanobubbles: The big picture. *Phys. A Stat. Mech. Appl.* **2002**, *314*, 696–705. [CrossRef]

23. Attard, P. The stability of nanobubbles. In *The European Physical Journal Special Topics*; Springer: Berlin/Heidelberg, Germany, 2013.
24. Zhang, X.; Chan, D.Y.; Wang, D.; Maeda, N. Stability of interfacial nanobubbles. *Langmuir* **2013**, *29*, 1017–1023. [CrossRef]
25. Cavalli, R.; Bisazza, A.; Giustetto, P.; Civra, A.; Lembo, D.; Trotta, G.; Guiot, C.; Trotta, M. Preparation and characterization of dextran nanobubbles for oxygen delivery. *Int. J. Pharm.* **2009**, *381*, 160–165. [CrossRef]
26. Wang, C.-H.; Huang, Y.-F.; Yeh, C.-K. Aptamer-conjugated nanobubbles for targeted ultrasound molecular imaging. *Langmuir* **2011**, *27*, 6971–6976. [CrossRef]
27. Xu, Q.; Nakajima, M.; Liu, Z.; Shiina, T. Biosurfactants for microbubble preparation and application. *Int. J. Mol. Sci.* **2011**, *12*, 462–475. [CrossRef]
28. Gerber, F.; Waton, G.; Krafft, M.P.; Vandamme, T.F. Long lived microbubbles for oxygen delivery. *Artif. Cells Blood Substit. Biotechnol.* **2007**, *35*, 119–124. [CrossRef]
29. Choi, D.; Heo, J.; Park, J.H.; Jo, Y.; Jeong, H.; Chang, M.; Choi, J.; Hong, J. Nano-film coatings onto collagen hydrogels with desired drug release. *J. Ind. Eng. Chem.* **2016**, *36*, 326–333. [CrossRef]
30. Khosravi-Darani, K.; Mozafari, M. Nanoliposome potentials in nanotherapy: A concise overview. *Int. J. Nanosci. Nanotechnol.* **2010**, *6*, 3–13.
31. Mozafari, M. Nanoliposomes: Preparation and analysis. In *Liposomes*; Springer: Berlin, Germany, 2010; pp. 29–50.
32. Unger, E.C.; Porter, T.; Culp, W.; Labell, R.; Matsunaga, T.; Zutshi, R. Therapeutic applications of lipid-coated microbubbles. *Adv. Drug Deliv. Rev.* **2004**, *56*, 1291–1314. [CrossRef]
33. Kheir, J.N.; Scharp, L.A.; Borden, M.A.; Swanson, E.J.; Loxley, A.; Reese, J.H.; Black, K.J.; Velazquez, L.A.; Thomson, L.M.; Walsh, B.K. Oxygen gas–filled microparticles provide intravenous oxygen delivery. *Sci. Transl. Med.* **2012**, *4*, 140ra188. [CrossRef]
34. Swanson, E.J.; Mohan, V.; Kheir, J.; Borden, M.A. Phospholipid-stabilized microbubble foam for injectable oxygen delivery. *Langmuir* **2010**, *26*, 15726–15729. [CrossRef]
35. Tsutsui, J.M.; Xie, F.; Porter, R.T. The use of microbubbles to target drug delivery. *Cardiovasc. Ultrasound* **2004**, *2*, 23. [CrossRef]
36. Harris, J.M.; Chess, R.B. Effect of pegylation on pharmaceuticals. *Nat. Rev. Drug Discov.* **2003**, *2*, 214. [CrossRef]
37. Torchilin, V.P. Recent advances with liposomes as pharmaceutical carriers. *Nat. Rev. Drug Discov.* **2005**, *4*, 145. [CrossRef]
38. Li, J.; Wang, X.; Zhang, T.; Wang, C.; Huang, Z.; Luo, X.; Deng, Y. A review on phospholipids and their main applications in drug delivery systems. *Asian J. Pharm. Sci.* **2015**, *10*, 81–98. [CrossRef]
39. Abou-Saleh, R.H.; Swain, M.; Evans, S.D.; Thomson, N.H. Poly (ethylene glycol) lipid-shelled microbubbles: Abundance, stability, and mechanical properties. *Langmuir* **2014**, *30*, 5557–5563. [CrossRef]
40. Nakamura, K.; Yamashita, K.; Itoh, Y.; Yoshino, K.; Nozawa, S.; Kasukawa, H. Comparative studies of polyethylene glycol-modified liposomes prepared using different peg-modification methods. *Biochim. Biophys. Acta Biomembr.* **2012**, *1818*, 2801–2807. [CrossRef]
41. Hwang, J.; Lee, E.; Kim, J.; Seo, Y.; Lee, K.H.; Hong, J.W.; Gilad, A.A.; Park, H.; Choi, J. Effective delivery of immunosuppressive drug molecules by silica coated iron oxide nanoparticles. *Colloids Surf. B Biointerfaces* **2016**, *142*, 290–296. [CrossRef]
42. Gajendiran, M.; Choi, J.; Kim, S.-J.; Kim, K.; Shin, H.; Koo, H.-J.; Kim, K. Conductive biomaterials for tissue engineering applications. *J. Ind. Eng. Chem.* **2017**, *51*, 12–26. [CrossRef]
43. Feshitan, J.A.; Legband, N.D.; Borden, M.A.; Terry, B.S. Systemic oxygen delivery by peritoneal perfusion of oxygen microbubbles. *Biomaterials* **2014**, *35*, 2600–2606. [CrossRef]
44. Hernot, S.; Klibanov, A.L. Microbubbles in ultrasound-triggered drug and gene delivery. *Adv. Drug Deliv. Rev.* **2008**, *60*, 1153–1166. [CrossRef]
45. Yoon, Y.I.; Kwon, Y.-S.; Cho, H.-S.; Heo, S.-H.; Park, K.S.; Park, S.G.; Lee, S.-H.; Hwang, S.I.; Kim, Y.I.; Jae, H.J. Ultrasound-mediated gene and drug delivery using a microbubble-liposome particle system. *Theranostics* **2014**, *4*, 1133. [CrossRef]
46. Swanson, E.J.; Borden, M.A. Injectable oxygen delivery based on protein-shelled microbubbles. *Nano Life* **2010**, *1*, 215–218. [CrossRef]

47. Polizzotti, B.D.; Thomson, L.M.; O'Connell, D.W.; McGowan, F.X.; Kheir, J.N. Optimization and characterization of stable lipid-based, oxygen-filled microbubbles by mixture design. *J. Biomed. Mater. Res. Part B Appl. Biomater.* **2014**, *102*, 1148–1156. [CrossRef]
48. Kheir, J.N.; Polizzotti, B.D.; Thomson, L.M.; O'Connell, D.W.; Black, K.J.; Lee, R.W.; Wilking, J.N.; Graham, A.C.; Bell, D.C.; McGowan, F.X. Bulk manufacture of concentrated oxygen gas-filled microparticles for intravenous oxygen delivery. *Adv. Healthc Mater.* **2013**, *2*, 1131–1141. [CrossRef]

nanomaterials

MDPI

Review

Depth-Sensing Indentation as a Micro- and Nanomechanical Approach to Characterisation of Mechanical Properties of Soft, Biological, and Biomimetic Materials

Nikolay V. Perepelkin [1,*], Feodor M. Borodich [1,2,*] and Alexander E. Kovalev [3] and Stanislav N. Gorb [3]

1 School of Engineering, Cardiff University, Cardiff CF24 3AA, UK
2 College of Aerospace Engineering, Chongqing University, Chongqing 400044, China
3 Department of Functional Morphology and Biomechanics, Zoological Institute of the University of Kiel, Kiel 24118, Germany; akovalev@zoologie.uni-kiel.de (A.E.K.); sgorb@zoologie.uni-kiel.de (S.N.G.)
* Correspondence: PerepelkinM@cardiff.ac.uk (N.V.P.); BorodichFM@cardiff.ac.uk (F.M.B.)

Received: 20 November 2019; Accepted: 14 December 2019; Published: 19 December 2019

Abstract: Classical methods of material testing become extremely complicated or impossible at micro-/nanoscale. At the same time, depth-sensing indentation (DSI) can be applied without much change at various length scales. However, interpretation of the DSI data needs to be done carefully, as length-scale dependent effects, such as adhesion, should be taken into account. This review paper is focused on different DSI approaches and factors that can lead to erroneous results, if conventional DSI methods are used for micro-/nanomechanical testing, or testing soft materials. We also review our recent advances in the development of a method that intrinsically takes adhesion effects in DSI into account: the Borodich–Galanov (BG) method, and its extended variant (eBG). The BG/eBG methods can be considered a framework made of the experimental part (DSI by means of spherical indenters), and the data processing part (data fitting based on the mathematical model of the experiment), with such distinctive features as intrinsic model-based account of adhesion, the ability to simultaneously estimate elastic and adhesive properties of materials, and non-destructive nature.

Keywords: characterization of materials; depth-sensing indentation; adhesion; the BG method; non-destructive testing

1. Introduction

Characterization of properties is a necessary, and one of the most important steps whenever a new material is created or discovered. Biological materials often demonstrate unique and outstanding properties in terms of structural strength, micro- and nanohierarchy, self-organizing and self-healing. Characterization of these properties is a challenging but important task, as properties discovered in biological specimens may lead to the development of new biomimetic and bioinspired materials and structures. The comprehensive review by Meyers et al. [1] provides insight on structure and properties of biological materials, and their various biomimetic counterparts.

Quite often, materials and structures change their mechanical properties in response to changing environment conditions. Changes in properties of biological tissues and cells can indicate various physiological states, including pathological ones. According to Wu et al. [2], "changes in cell and nuclear mechanics are hallmarks of many human diseases, particularly metastatic cancer, cardiovascular disease, inflammation, laminopathies, host-microbe interactions in infectious diseases, and frailty in aging".

Despite its practical importance, characterization of mechanical properties of soft materials, including biological ones, is a challenging task. Classical methods of material characterization are well developed and standardized for various purposes, such as the determination of Young's modulus [3,4], Poisson's ratio [5] or adhesive strength [6]. However, carrying out classical materials testing becomes increasingly difficult when the tested material is very soft, or is available in small quantities. Kang et al. [7] argue that "testing of micro/nanoscale material specimens shares many similarities with conventional, standardized methods... However, as specimen size decreases, testing presents additional challenges... For example, decreasing sample size results in critical challenges in the preparation, handling, and gripping of specimens; in the application of small displacements and forces; in high resolution stress and strain measurements; and in precise control." The above statement is confirmed by the real-time materials testing inside scanning electron microscopes via nanomanipulation.

For instance, Cao et al. [8,9] carried out nanomechanical tensile testing of a graphene oxide nanosheets using a microelectromechanical system (MEMS) located directly inside an electron microscope. Clearly, it was impossible to create graphene oxide nanospecimens of pre-defined shape. Hence, to interpret the experimental measurements and estimate stress distribution in the specimens, Cao et al. [8] employed molecular dynamics and Finite Element Method (FEM) simulations that took into account the actual shape and thickness of the graphene specimens. The latter papers clearly demonstrate that size effects makes it increasingly difficult to follow conventional macroscopic experimental procedures.

Nevertheless, such experimental technique as depth-sensing indentation, can be applied without much change to various kinds of materials, including soft materials, and at various length scales. Cantilever-based mechanical probing, such as atomic force microscopy, becomes the only available method for elasticity mapping of living specimens at nanometre length scale [10].

Very often biological structures have gradient of material properties in different directions (length, width, depth) down to microscale [11,12]. This fact makes it impossible to use any kinds of tensile tests for characterisation of gradual changes of material properties. At the same time, various kinds of DSI (micro-/nanoindenter, atomic force microscope) allow testing of such biological materials at various length scales, from macroscale [13] up to nanoscale [14].

It is important to note that when it comes to mechanical contact of elastic bodies, both elastic and adhesive properties of contacting materials affect the resulting measurements. Thus, despite DSI itself is done in similar way at various length scales, the interpretation of DSI data should take into account length scale effects, such as adhesion, in many important situations.

We also need to underline the importance of surface roughness and consideration of related changes in real contact area that can affect adhesion. Experiments of Purtov et al. [15] and modelling of the experiments by Pepelyshev et al. [16], employing Galanov's model of adhesion between rough surfaces [17], showed that consideration of real contact area is a very important factor in adhesive contact problems. This statement is also nicely confirmed by the analysis of nanoscale static friction by Polyakov et al. [18], and various works that demonstrate the influence of roughness on the value of adhesive pull-off force (e.g., [19–21]).

Adhesive phenomena become increasingly important, as the size of the contact area decreases. Kendall [22] described the nano-scale world as "he sticky universe", emphasizing the domination of adhesive interactions, such as van der Waals forces, at this length scale. Inability to take adhesion into account may also lead to erroneous results when DSI of soft specimens is carried out. Kohn and Ebenstein [23] argue that "although adhesion leads to overestimation of the modulus of compliant samples when analyzing nanoindentation data using traditional analysis techniques, most studies of biomaterials have ignored its effects ...Compliant, hydrated materials exhibit very different mechanical behaviors relative to mineralized tissues, and require significant modifications to traditional indentation methods to measure accurate modulus values". It becomes clear that methods of DSI of soft materials and DSI at micro-/nanoscale should take adhesion into account intrinsically and should be based on mathematical models of adhesive contact, rather than non-adhesive ones. It is worth noting that viscoelasticity, which is inherent in many biological materials [24,25], also makes it difficult to use classical contact mechanics models for the estimation of properties by DSI. However, viscoelastic phenomena are not considered in this work.

In this review paper, we focus the discussion on different DSI approaches, their critical assessment, and our recent advances related to the development of a method for simultaneous identification of elastic and adhesive properties of materials and structures based on experimental measurements of DSI: the BG method, and its extended variant: the eBG method. The BG/eBG methods serve the same purpose in DSI with adhesion, as the well-known Oliver–Pharr method [26] in the conventional non-adhesive instrumental indentation: to provide the complete framework that includes experimental technique, data processing algorithm, and the verification of the model used for data processing. As long as the BG method takes into account adhesive effects in DSI, its scope extends beyond the scope of the Oliver–Pharr method, in particular, to the problems of indentation of soft materials, including biological ones, and indentation at micro- and nanoscale.

The paper is organized as follows. In Section 2, conventional DSI techniques are discussed and critically assessed: (i) identification of elastic properties of specimens using sharp pyramidal indenters, and the Oliver-Pharr method in particular; (ii) identification of adhesive properties using pull-off indentation tests by means of spherical indenters. Sections 3 and 4 are devoted to the description of the BG and eBG methods, correspondingly, including scope, workflow, advantages and limitations. A numerical example based on the use of the eBG method and an asymptotic model of adhesive indentation of a thin elastic coating is presented in Section 4 as well.

2. Depth-Sensing Indentation: Conventional Approaches

In this section, we discuss commonly used approaches and techniques of DSI, including critical reviews. In particular, the importance of adhesion in DSI of soft materials is emphasized.

DSI combined with mathematical methods of contact mechanics allows one to determine such properties of the tested materials as reduced elastic contact modulus E^* and the work of adhesion w. The modulus E^* is defined as

$$(E^*)^{-1} = (1-\nu^2)E^{-1} + (1-\nu_i^2)E_i^{-1}, \quad (1)$$

where the elastic moduli and Poisson's ratios of the specimen and the indenter are denoted as E, ν, and E_i and ν_i respectively. The quantity E^* arises in the Hertz contact theory as the proportionality coefficient, among others, between contact pressure and the total approach of two contacting elastic half-spaces assumed to have paraboloidal shapes [27]. Quite often it is possible to assume the indenter to be a rigid body ($E_i = \infty$), hence E^* depends only on the properties of the specimen $E^* = E/(1 - \nu^2)$). The work of adhesion w can be defined as the amount of work, per unit area, required to separate two contacting surfaces to infinite distance [27]. See also discussion in [28] related to the thermodynamic interpretation of this quantity. Practical ways to experimentally estimate the work of adhesion are considered further in the paper.

In commonly used DSI approaches these two quantities, E^* and w, are identified independently, using two different tests involving nominally sharp pyramidal and smooth spherical indenters respectively. These techniques are discussed below.

2.1. DSI by Sharp Indenters: The BASh Formula and the Oliver–Pharr Approach

To the best of our knowledge, Kalei [29] was the first researcher who introduced DSI as an experimental technique. In the early years of the existence of this method, it was common to use sharp pyramidal indenters, e.g., developed by M. Khrushchov and E. Berkovich [30] and later named "Berkovich indenters".

Although we use the term "sharp indenter", the reader should realise that nominally sharp indenters (at macroscale) are blunt at nanoscale. For example, Borodich et al. [31] argued that the the indenter shape near the tip can be well approximated by a monomial (power-law) functions of radius. If degree of the monomial $d = 1$ then this is mathematically sharp indenter (cone or pyramid), while if the degree $d = 2$ then this may be interpreted as a sphere or an elliptical paraboloid. They demonstrated that shape of real indenters they considered may vary between 1 and 2. However, more often the degree d is close to 2 (see e.g., Kindrachuk et al. [32]), hence it is customary to describe the indenter bluntness by the radius of its spherical approximation. One has to realise that the tip angle of a nanoindenter is much greater than the angle of an AFM tip, therefore the geometrically linear formulation of contact problem may be applied to AFM tests only for very shallow indentation depth.

Shortly after the invention of DSI, it would become clear that the unloading branch of the force-displacement curve (the $P - \delta$ curve) could be approximately described using the classical Hertz contact theory. Bulychev et al. [33] suggested to modify known exact formulae of axisymmetric contact for that purpose. In particular, they noticed that the slope of the force-displacement curve for a spherical indenter satisfies the relation

$$S = \frac{dP}{d\delta} = 2E^* a \tag{2}$$

where a is the contact radius.

The expression (2) was then re-written in the following form that is often referred to as the Bulychev–Alekhin–Shorshorov (BASh) relation [34]):

$$S = \frac{dP}{d\delta} = 2E^* \sqrt{A/\pi}. \tag{3}$$

In the latter formula, the expression $\sqrt{A/\pi}$ replaces the contact radius a (A is the current contact area). The resulting formula therefore becomes an approximate relation, rather than the exact one (2). However, this allowed material scientists community to apply the BASh relation (3) to the case of widely used pyramidal indenters. The latter fact made the expression (3) the corner stone of commonly used techniques of instrumental indentation, including the one described by Oliver and Pharr (OP) [26], which has become de facto the industrial standard [35].

Indeed, the approximate nature of formula (3) makes it complicate to properly process and interpret indentation results. For instance, Oliver and Pharr [26] had to use rather complex approximate relation between the contact area and the indentation depth δ

$$A = 24.5\delta^2 + C_0\delta + C_1\delta^{1/2} + C_2\delta^{1/4} + C_3\delta^{1/8} + \ldots + C_n\delta^{1/2^n} \tag{4}$$

where "the lead term describes a perfect Berkovich indenter; the others describe deviations from the Berkovich geometry due to blunting at the tip" [26]. Complications also arise from the "pile-up" or "sink-in" of the material around the indenter [36]. Critical discussion related to the OP method can be found in a number of papers, such as [37–40].

In the recent years, possible improvements to the BASh formula have been suggested by means of correction factors, e.g., [41]:

$$\frac{dP}{d\delta} = \beta_1\beta_2\beta_3 \frac{2}{\sqrt{\pi}} E^* \sqrt{A}, \tag{5}$$

where $\beta_1 \ldots \beta_3$ are correction factors taking into account various aspects of elastoplastic deformation and friction in the contact zone between the indenter and the specimen. The factor β_1 is introduced due to the concept of the effective indenter shape (the Galanov effect) [42,43], β_2 is the factor related to non-circular shape of the indenter imprint, and the factor β_3 is introduced due to the effects of friction between the indenter and the specimen [37,44]. In a recent work, however, Galanov and Dub [39] showed that it is not correct to take the Galanov effect into account by simply using the correction factor β_1. They argued that neglecting the true distance between the indenter and the deformed surface of the specimen lead to wrong estimation in the problem governing parameter δ (indentation depth). Therefore, straightforward application of the OP approach may lead to certain inaccuracies in estimations of the reduced contact modulus E^*.

Previously, we tested properties of insect adhesive structures [11], insect joints [12], insect cuticle [45,46], snake skin [47,48], gecko setae [49], human teeth [50], mollusk teeth [51], and plant materials [45] using nanoindentation by either Berkovich indenter or AFM tip. However, from practical point of view, the use of nominally sharp indenters sometimes may not be desired, e.g., due to the destructive nature of such a method. What is even more important, is that the above technique does not take adhesion into account which may lead to errors when nanoscale measurements are considered.

2.2. Adhesion in Depth-Sensing Indentation, and the Conventional Use of Spherical Indenters

The presence of adhesion in the DSI experiment is twofold. The first aspect is that DSI can be specifically used to study adhesion of different materials [52–56]. Previously, we tested mechanical properties of insect adhesive pads [24,57] and other insect structures containing rubber-like protein resilin [25] using microindentation by spherical indenter tip. It is also very useful approach to measure adhesive properties of biological [58] and biomimetic [59,60] materials. The classic theories of adhesive frictionless contact, including the Johnson-Kendall-Roberts (JKR) [61], Derjaguin-Muller-Toporov (DMT) [62] and Maugis (see e.g., [28]) theories, are well developed and can be readily used for the above purposes.

Most often, the identification of the work of adhesion w is based on the use of direct methods [54,55]. One of the most popular approaches is based on the JKR model and direct experimental measurements of the adherence pull-off force P_{adh} between a sphere of radius R and an a specimen considered an elastic half-space. The JKR theory readily provides relation between the work of adhesion w and P_{adh} in the following form:

$$w = -\frac{2}{3}\frac{P_{adh}}{\pi R}. \tag{6}$$

However, there is evidence that direct measurement of the pull-off force may have poor reproducibility, as the tensile (adhesive) part of the force-displacement readings may be influenced by surface quality. Among the factors that influence pull-off force readings researchers point out surface roughness of the specimen [21] (see also [63]), quality of the probe, e.g., wear, the presence of adsorbed chemicals on the probe surface. These factors become increasingly important at nanoscale, in case of DSI done by means of atomic force microscopy (AFM). In such conditions, one needs to carry out a large number of tests to estimate w properly using (6) or a similar method.

The other reason to consider DSI an adhesive contact problem is that adhesion greatly influences the results of the DSI experiment at micro-/nanoscale [64], or when it comes to the investigation of properties of soft and biological materials. Disregarding adhesion in these experiments may lead to significant errors in experimental results. Other researchers support this point too, e.g., Kohn and Ebenstein [23] quoted earlier in the Introduction. While some researchers [23,65] develop modifications to the conventional DSI methods involving sharp indenters, it looks more natural to build the complete methodology around an approach that takes adhesion into account intrinsically. In the following sections we describe a method based on this core concept: the BG method, and its extended variant.

3. The BG Method

In 2008, Borodich and Galanov developed a new method (the BG method [64,66–68]), which constituted a two-stage framework: (a) experimental stage (DSI by spherical indenters), and (b) data processing stage (fitting the experimental data with an appropriate mathematical model of the DSI test with subsequent verification). Such a workflow allows one to intrinsically take adhesion into account, and evaluate both elastic and adhesive properties of the specimen from a single experiment, not two separate ones, as it is done in conventional DSI discussed above. In this Section we consider the scope and detailed workflow of the original BG method, while its extensions are considered further in the text.

The scope of the original BG method is simultaneous determination of the reduced elastic modulus E^* and the work of adhesion w in the framework of the classical theories of adhesive contact, like the JKR or DMT. In order to apply these theories, the specimen is considered an elastic half-space.

The workflow of the method begins with depth-sensing indentation by a spherical indenter which results in N measurements of the indentation depth δ_i and indentation force P_i during DSI: (δ_i, P_i), $i = 1 \ldots N$. The basic idea of the method is to find the unknown material properties by taking a mathematical model of the same experiment, that is, the equation that theoretically describes the force-displacement relation, and adjusting control parameters of the model so that the theoretical force-displacement curve best fits the data in certain sense.

In the beginning, the force-displacement dependency is written in a dimensionless form. For example, the governing equations of both the JKR and DMT theories can be represented as

$$F\left(\frac{P}{P_c}, \frac{\delta}{\delta_c}\right) = 0. \tag{7}$$

Note the characteristic scales (scaling parameters) of the problem P_c and δ_c. They are introduced to formulate the force-displacement curve in the dimensionless form. In the BG method, these quantities are subject to adjustment, while the theoretical force-displacement curve (7) is fit to the experimental data points as the result of an optimization process.

Technically, to determine two unknown quantities, P_c and δ_c, one would need only two equations. Hence, if all the experimental readings were exact, the following identities would be all true at the same time

$$F\left(\frac{P_i}{P_c}, \frac{\delta_i}{\delta_c}\right) = 0, \qquad i = 1, \ldots, N. \tag{8}$$

In that case, one would just need to select any two of them and solve for P_c and δ_c.

In the real experiment, however, measuremets (δ_i, P_i) always contain certain inaccuracies. Therefore, in general case, the relations (8) are never true all at the same time.

However, one may consider (8) an overdetermined system of equations. Instead of making all the expressions exact at the same time, it is possible to minimize the measure of the total error produced in the system. Consider the quantity

$$\varepsilon_i = F\left(\frac{P_i}{P_c}, \frac{\delta_i}{\delta_c}\right) \tag{9}$$

which is the individual residual of i-th equation. If one minimizes the mean square residual $e = \frac{1}{N}\sum_{i=1}^{N} \varepsilon_i^2$ of the overdetermined system (8), it necessarily leads to the following optimization problem:

$$\{P_c^*, \delta_c^*\} = \arg\min \Phi(P_c, \delta_c) \tag{10}$$

$$\Phi(P_c, \delta_c) = \sum_{i=1}^{N} \left[F\left(\frac{P_i}{P_c}, \frac{\delta_i}{\delta_c}\right)\right]^2. \tag{11}$$

Here P_c^*, δ_c^* are the optimal values of the characteristic parameters that minimize the objective functional Φ and provide best fit of the experimental data. Clearly, the objective functional is the mean square residual multiplied by the number of data points N, which is constant.

Methods of solution of such problems are well developed (see e.g., [69,70]). As long as the optimal values of the characteristic parameters $P_c = P_c^*$ and $\delta_c = \delta_c^*$ are found, the sought material parameters

E^* and w can be readily evaluated from them, because the characteristic parameters are known expressions that are defined within the framework of the chosen theory of adhesive contact.

The particular representation of the theoretical curve (7) and the characteristic scales can be different, dependidng on the theory of adhesive contact used to describe the DSI test. For example, in the JKR theory, the characteristic scales may be taken as

$$P_c = \frac{3}{2}\pi w R,\ \delta_c = \frac{3}{4}\left(\frac{\pi^2 w^2 R}{E^{*2}}\right)^{1/3} \tag{12}$$

when a spherical indenter of radius R is considered. The particular look of the theoretical force-displacement curve (7) in this case is

$$F\left(\frac{P}{P_c}, \frac{\delta}{\delta_c}\right) = \begin{cases} (3\chi - 1)\left((1+\chi)/9\right)^{\frac{1}{3}} - \delta/\delta_c = 0, & (\chi \geq 0,\ \delta/\delta_c \geq -3^{-2/3}), \\ (3\chi + 1)\left((1-\chi)/9\right)^{\frac{1}{3}} - \delta/\delta_c = 0, & (2/3 \geq \chi \geq 0,\ -3^{-2/3} > \delta/\delta_c \geq -1) \end{cases} \tag{13}$$

where $\chi = \sqrt{1 + P/P_c}$ [28].

In the JKR theory, P_c defines the maximum of the absolute value of the pull-off force, and δ_c is the absolute value of minimum possible displacement among all points representing the $P - \delta$ curve. The material properties E^* and w can be found from Equation (12) as

$$w = \frac{2P_c}{3\pi R}, \qquad E^* = \frac{P_c}{4}\sqrt{\frac{3}{R\delta_c^3}}. \tag{14}$$

The DMT theory can be used in the BG method too [67]. In the DMT theory, the force-displacement curve can be written in the form [28]

$$F\left(\frac{P}{P_c}, \frac{\delta}{\delta_c}\right) = \frac{P}{P_c} - \frac{1}{\sqrt{3}}\left(\frac{\delta}{\delta_c}\right)^{3/2} + \frac{4}{3} = 0 \tag{15}$$

with the characteristic parameters defined above.

For the case of the DMT theory, Borodich et al. [67] showed that the evaluation of P_c and δ_c can be separated from one another by presenting the corresponding overdetermined system (8) as

$$3^{1/3}\left(\frac{P_i}{P_c} + \frac{4}{3}\right)^{2/3} - \frac{\delta_i}{\delta_c} = 0, \qquad i = 1, \ldots, N, \tag{16}$$

and re-writing it in the form

$$a_i P_c - b_i = 0, \tag{17}$$

$$c_i(P_c)\delta_c - \delta_i = 0, \tag{18}$$

where expressions a_i and b_i solely depend on the experimental data, and the expression c_i depends on both experimental data and P_c.

The final step of processing experimental data in the BG method is the validation of the initially chosen theory of adhesive contact, and the corresponding mathematical model (7). Clearly, a given data set can be fitted with multiple theoretical force-displacement curves that belong to different theories of adhesive contact. For this reason, the BG method has a verification step: calculation of the Tabor–Muller parameter introduced in the works of Tabor [71] and Muller [72] (see also discussion in [28]). The purpose of this dimensional parameter is to provide clear distinction of applicability range between the JKR and DMT theories of adhesive contact:

$$\mu = \left(\frac{Rw^2}{E^{*2} z_0^3} \right)^{1/3}. \quad (19)$$

Here, R is the effective curvature radius of contacting bodies, and z_0 is the equilibrium distance between atoms of the contacting bodies, normally assumed to be 0.3...0.5 nm. The case $\mu \gg 1$ is normally referred to the JKR theory, while values $\mu \ll 1$ suggest to use the DMT theory. If the calculated value of μ does not correspond to the theory of adhesive contact that was used in the above calculations, then a different theory should be chosen, and data fitting repeated.

Note that other authors have also used data fitting-related techniques in studies of soft and biological materials. See, for instance, the works [73–78]. However, in most cases one can hardly find any detailed description of fitting methods and procedures. Performance-related aspects in terms of accuracy and robustness are omitted too. Only few authors (e.g., [79–81]) devote the whole study to the development of methodology of identification of material properties. With this regard, we should note that the above mentioned works [64,66–68] contain detailed analysis of the mathematical aspects of the BG method, and its accuracy, including the ability to withstand the presence of measurement noise. The recent work by Perepelkin et al. [82] contains detailed discussion and performance assessment with regards to the extended BG method, which is discussed in the next Section.

4. The Extended BG (eBG) Method

In contact mechanics of adhesive contact, only solutions of the simplest, classical contact theories, like JKR or DMT, can be represented as explicit functions. Overall, it is typical to represent the solutions of more complex contact problems in the form of parametric functions, in which both the external load and the indenter displacement depend on the contact radius as the parameter [34]. Among numerous examples of this kind, one can find asymptotic mathematical models of JKR-type contact for layered and coated medium [83–86], implementations of the Maugis theory [87,88], or the double-Hertz theory [89]. These models have complex mathematical form of parametric functions that cannot be exactly reduced to explicit or implicit ones. In addition, it is fundamentally impossible to do such reduction to semi-analytical models based on the finite elements method calculations [90–92].

In its initial formulation, the BG method cannot utilize mathematical models which describe force-displacement curve as a parametric function. The focus of the original method is on the use of the classical theories of adhesive contact. However, this issue has recently been addressed, as Perepelkin et al. [93] announced the eBG method, an extended version of the original approach which has distinctive new features, such as orthogonal distance fitting concept, and two stage fitting approach. Theoretical aspects of the eBG method are presented in the recent work [82], which describes two different implementations of the eBG method. Here we briefly discuss the variant of the eBG method which was successfully used in [93].

The scope of the eBG method is determination of elastic moduli (such as reduced contact modulus E^*) and the work of adhesion w of elastic structures, for which the force-displacement DSI curve can be described as a parametric function:

$$\begin{cases} \delta = \delta_c f_1(\bar{a}, \delta_c, P_c), \\ P = P_c f_2(\bar{a}, \delta_c, P_c). \end{cases} \quad (20)$$

In these expressions, $\bar{a}$ is the dimensionless parameter, which may have physical interpretation, e.g., dimensionless contact radius.

Normally, fitting a parametric curve requires one to find the values $\bar{a}_i$ corresponding to each data point from the experimental data (δ_i, P_i), $i = 1 \ldots N$. In turn, this increases the number of unknown quantities to the order of thousands, whilst only two quantities, P_c and δ_c, are the matter of interest for the researcher. To avoid this issue, we introduced [82,93] a two-stage data fitting process.

The first stage can be considered data filtering, as typically experimental data contains measurement noise. It is supposed to first smooth the experimental data by fitting it with some auxiliary curve $P = \Psi(\delta)$ (Figure 1). To make fitting relatively fast and simple process, the mathematical form of $\Psi(\delta)$ should be rather simple.

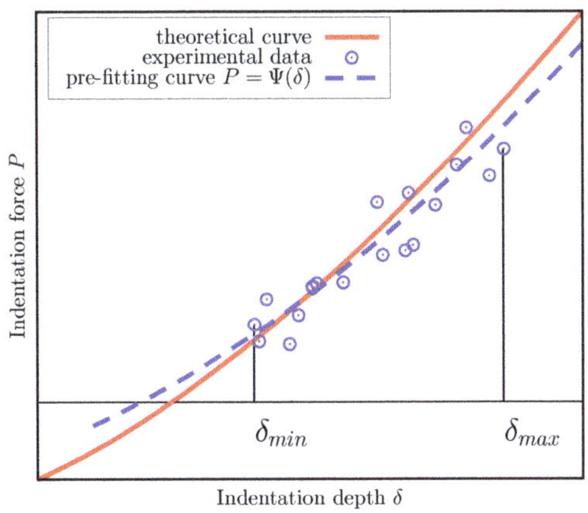

Figure 1. Preliminary fitting the experimental data with an auxiliary curve $\Psi(\delta)$.

If the auxiliary fitting curve (the pre-fitting curve) has simple mathematical form, then advanced fitting techniques can be implemented. In our works, we built such auxiliary fitting curves that best fit experimental data in terms of minimal averaged value of squared orthogonal distance from the data points to the curve (the orthogonal distance curve fitting (ODF) approach [94,95]). The benefit of this approach is that it filters out noise present in both experimental measurements, δ_i and P_i, whereas the commonly used least squares fitting minimizes residual related to only one measured quantity, either δ_i or P_i.

Orthogonal distance curve fitting problems are known to be computationally intensive [94,95], hence preliminary fitting (pre-fitting) was done using a polygonal chain (a piecewise-linear function) with the number of segments N_S, which could vary.

On the second stage the values of P_c and δ_c become subject to adjustment, so that the theoretical curve becomes closest to the auxiliary one. This is achieved by minimizing the squared norm of difference of the two functions:

$$\hat{\Phi}(P_c, \delta_c) = \int_{\delta_{min}}^{\delta_{max}} [P(\delta) - \Psi(\delta)]^2 \, d\delta. \tag{21}$$

$$\{P_c^*, \delta_c^*\} = \arg\min \hat{\Phi}(P_c, \delta_c) \tag{22}$$

The optimal values of the scaling parameters, P_c^*, δ_c^*, are further used to evaluate unknown material properties as discussed in the previous section.

The actual force-displacement curve is supposed to have parametric form (20), therefore, substitution of (20) yields the final form of the objective functional of the eBG method $\hat{\Phi}$:

$$\hat{\Phi}(P_c, \delta_c) = \delta_c \int_{\bar{a}_{min}}^{\bar{a}_{max}} [P_c f_2(\bar{a}, \delta_c, P_c) - \Psi(\delta_c f_1(\bar{a}, \delta_c, P_c))]^2 \frac{\partial f_1(\bar{a}, \delta_c, P_c)}{\partial \bar{a}} d\bar{a}. \tag{23}$$

In this functional, $\bar{a}$ is the integration variable. Hence, the optimization problem has dimension two, and the need to find the corresponding value of the parameter $\bar{a}$ for every data point is therefore eliminated.

The above two-stage approach has been experimentally validated in [93] by conducting two independent experiments: (a) a DSI experiment with subsequent use of the eBG method, and (b) a tensile testing of dumbbell specimens made of the same material. The values of the reduced elastic modulus obtained form DSI data by means of the eBG method, and the ones obtained from the tensile testing were in good agreement.

The work [82] contains a number of numerical benchmarks related to the speed and accuracy of the variant of the eBG method shown above. In this paper, however, we present a short numerical example, in which DSI of a thin elastic layer bonded to a rigid substrate is simulated (Figure 2). The indenter is supposed to be a sphere of a known radius R.

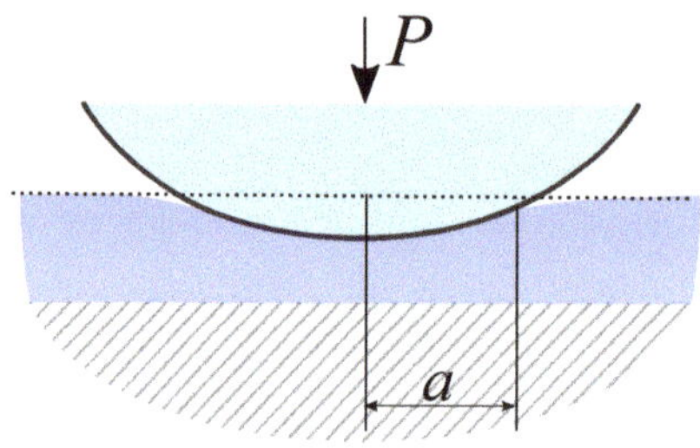

Figure 2. Indentation of a thin elastic layer bonded to the rigid base.

The mathematical model of such a DSI experiment in the framework of the JKR theory of adhesive contact can be written in a dimensionless form [85]:

$$\frac{P}{P_c} = \bar{a}^4 - 2\bar{a}^2, \qquad \frac{\delta}{\delta_c} = \bar{a}^2 - 1, \tag{24}$$

where the characteristic parameters are

$$P_c = 2\pi R w, \quad \delta_c = \sqrt{\frac{2w}{K}}, \tag{25}$$

and the model parameters are as follows: $K = E\,(1-\nu)/[h\,(1+\nu)\,(1-2\nu)]$ is the stiffness of an equivalent elastic foundation substituting the layer, h is the layer thickness, E and ν are the Young's modulus and Poisson's ratio of the layer, respectively. The indenter is supposed to be rigid, therefore the reduced contact modulus of the layer is $E^* = E/(1-\nu^2)$. The simulation workflow is shown in Figure 3.

The simulated model parameters were: $E = 0.91$ MPa, $\nu = 0.3$, that is $E^* = 1$ MPa; $R = 10$ mm, $h = 0.1$ mm, $w = 100$ mJ/m^2. In the simulation, the Tabor–Muller parameter was calculated after setting all the parameters. Hence, the validity of the JKR theory in this scenario was confirmed a priori, and no further checks were done. The theoretical force-displacement curve is depicted if Figure 4 as the solid line. The compressive part of it was used to generate noisy data emulating a DSI experiment. The noisy data is denoted by points in Figure 4 (shifted).

The values of the indentation depth δ were shifted by a random quantity emulating the uncertainty in the determination of the true indenter position, as the exact distance from the indenter to the point of contact is never known in the real experiment. This shift value was determined as follows. A series of possible shift values was generated. For each shift value minimization of (23) was performed using the modified data set, as the shift value was subtracted from the initially given values δ_i $(i = 1, \ldots, N)$. The absolute minimum of values of the objective functional among all trial minimizations was identified, and the corresponding shift value was supposed to be the true one. Further, the values of P_c and δ_c found in the corresponding trial problem were supposed to be the true ones. The issue of uncertainty in the indenter position is discussed in detail in [82].

Two cases were considered in the simulation: (i) high noise scenario (Figure 4a), and (ii) low noise scenario (Figure 4b). In each scenario 20 data sets were generated using normally distributed noise. The eBG method, as shown above, was applied to each data set separately. In each individual case, data points were pre-fitted with a seven-segment polygonal chain using the orthogonal distance fitting approach.

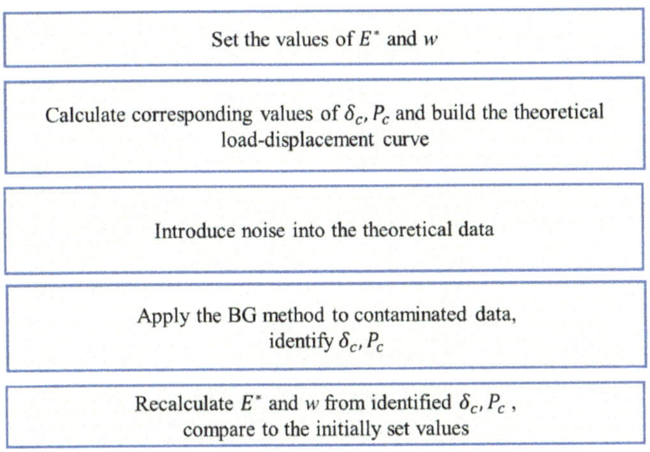

Figure 3. The workflow of the numerical simulation demonstrating the accuracy and robustness of the extended Borodich–Galanov (eBG) method.

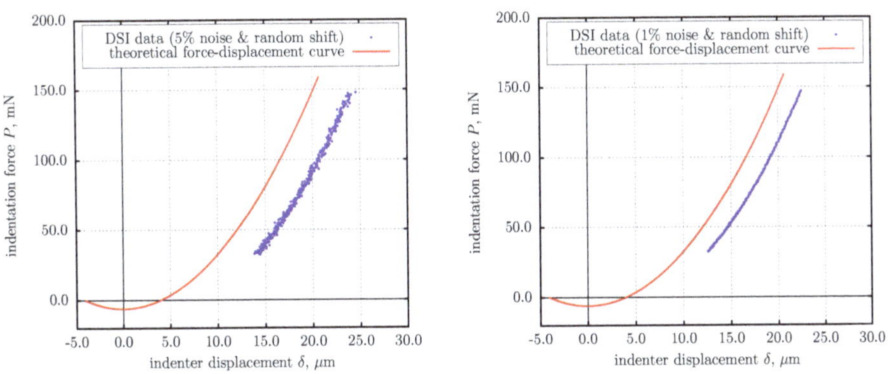

(a) (b)

Figure 4. Numerical simulation. The theoretical force-displacement curve (solid line) and an example data sets simulating DSI readings (dots, shifted). (**a**) High noise scenario. (**b**) Low noise scenario.

The individual identified values of E^* and w are depicted graphically in Figure 5. It is remarkable, that despite significant spread in the individual identified values, the averaged values of E^* and w

were determined with high accuracy: $E^* = 0.9948$ MPa and $w = 99.38$ mJ/m^2 in the high noise scenario, $E^* = 0.9987$ MPa and $w = 101.2$ mJ/m^2 in the low noise scenario.

This example demonstrates an important feature of the BG/eBG method: the ability to extract unknown material properties, including adhesive ones, from the stable compressive part of the force-displacement data, in contrast to the methods based on the measurement of the pull-off force, which use less stable tensile part of the DSI data.

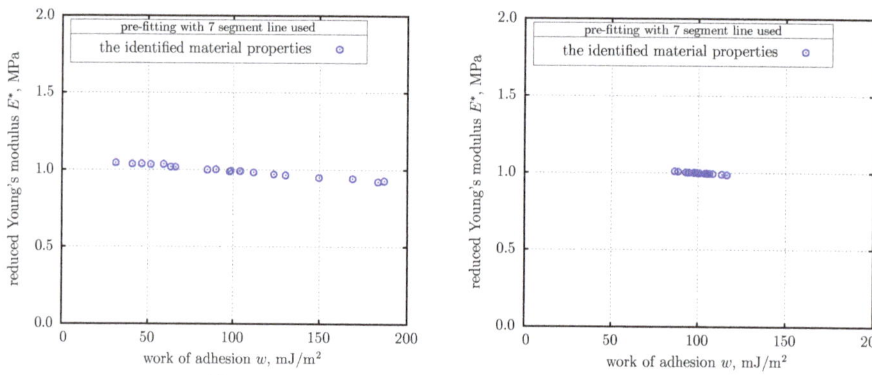

(a) (b)

Figure 5. Numerical simulation. The results of identification of material properties from 20 data sets containing noise and random coordinate origin shift. (**a**) High noise scenario. (**b**) Low noise scenario.

5. Conclusions

As specimen size decreases, classical methods of material testing become extremely complicated and challenging. At the same time, such experimental technique as depth-sensing indentation (DSI) can be applied without much change at macro-, micro-, and nanoscale. Interpretation of the DSI data needs to be done carefully though, as length-scale dependent effects, such as adhesion, should be taken into account.

In this work, we review our recent advances in the development of a method that intrinsically takes adhesion in DSI into account: the BG method, and its extended variant, the eBG method. The BG/eBG methods can be considered a framework made of the experimental part (DSI by means of spherical indenters), and the data processing part (data fitting based on the mathematical model of the experiment).

The main advantages and distinctive features of the BG/eBG methods are: (i) intrinsic account of adhesion, in contrast to the common methods of instrumental indentation (e.g., the Oliver–Pharr method). Hence, they can be used for determination of adhesive properties in the first instance, or whenever adhesion during DSI cannot be neglected (i.e., micro/nanoindentation, and indentation of soft solids); (ii) the ability to use only stable compressive DSI data to identify unknown material properties, including adhesive ones; (iii) the ability to estimate both elastic and adhesive properties of the specimen in a single experiment, instead of two separate ones, as it is usually done; (iv) the BG method is a non-destructive technique. This provides benefits in terms of: (a) the ability to test living biological samples (e.g., cell membranes); (b) the ability to do repetitive tests at exactly the same location (e.g., to do statistical averaging of the results, or to investigate surface physics under changing environment conditions); (c) non-destructive control of the quality and properties of protective coatings.

Funding: Nikolay Perepelkin acknowledges that his participation in this project became possible due to funding from the European Union's Horizon 2020 research and innovation programme under the Marie Skłodowska-Curie grant agreement No 663830.

Acknowledgments: Feodor Borodich gratefully acknowledges that his visits to the Functional Morphology and Biomechanics Group at Kiel University were supported by Alexander von Humboldt Foundation, Germany.

Conflicts of Interest: The authors declare no conflict of interest.

References

1. Meyers, M.A.; Chen, P.Y.; Lin, A.Y.M.; Seki, Y. Biological materials: Structure and mechanical properties. *Progress Mater. Sci.* **2008**, *53*, 1–206. [CrossRef]
2. Wu, P.H.; Aroush, D.R.B.; Asnacios, A.; Chen, W.C.; Dokukin, M.E.; Doss, B.L.; Durand-Smet, P.; Ekpenyong, A.; Guck, J.; Guz, N.V.; et al. A comparison of methods to assess cell mechanical properties. *Nat. Methods* **2018**, *15*, 491–498. [CrossRef] [PubMed]
3. ASTM. *International Standard ASTM E111-17. Standard Test Method for Young's Modulus, Tangent Modulus, and Chord Modulus*; ASTM E111-17; ASTM: West Conshohocken, PA, USA, 2017.
4. International Organization for Standardization. *International Standard ISO 527-1. Plastics—Determination of Tensile Properties—Part 1: General Principles*; ISO527-1; ISO: Geneva, Switzerland, 2012.
5. ASTM. *International Standard ASTM E132-17. Standard Test Method for Poisson's Ratio at Room Temperature*; ASTM E132-17; ASTM: West Conshohocken, PA, USA, 2017.
6. International Organization for Standardization. *International Standard ISO 29862. Self Adhesive Tapes—Determination of Peel Adhesion Properties*; ISO 29862; ISO: Geneva, Switzerland, 2018.
7. Kang, W.; Merrill, M.; Wheeler, J.M. In situ thermomechanical testing methods for micro/nano-scale materials. *Nanoscale* **2017**, *9*, 2666–2688. [CrossRef] [PubMed]
8. Cao, C.; Mukherjee, S.; Howe, J.Y.; Perovic, D.D.; Sun, Y.; Singh, C.V.; Filleter, T. Nonlinear fracture toughness measurement and crack propagation resistance of functionalized graphene multilayers. *Sci. Adv.* **2018**, *4*, eaao7202. [CrossRef] [PubMed]
9. Cao, C.; Howe, J.Y.; Perovic, D.; Filleter, T.; Sun, Y. In situ TEM tensile testing of carbon-linked graphene oxide nanosheets using a MEMS device. *Nanotechnology* **2016**, *27*, 28LT01. [CrossRef] [PubMed]
10. Alekya, B.; Rao, S.; Pandya, H.J. Engineering approaches for characterizing soft tissue mechanical properties: A review. *Clin. Biomech.* **2019**, *69*, 127–140.
11. Peisker, H.; Michels, J.; Gorb, S. Evidence for a material gradient in the adhesive tarsal setae of the ladybird beetle *Coccinella septempunctata*. *Nat. Commun.* **2013**, *4*, 1–7. [CrossRef]
12. Barbakadze, N.; Enders, S.; Gorb, S.; Arzt, E. Local mechanical properties of the head articulation cuticle in the beetle *Pachnoda marginata* (*Coleoptera, Scarabaeidae*). *J. Exp. Biol.* **2006**, *209*, 722–730. [CrossRef]
13. Griffin, M.; Premakumar, Y.; Seifalian, A.; Butler, P.E.; Szarko, M. Biomechanical Characterization of Human Soft Tissues Using Indentation and Tensile Testing. *J. Vis. Exp.* **2016**, *118*, 54872. [CrossRef]
14. Bhushan, B.; Li, X. Nanomechanical characterisation of solid surfaces and thin films. *Int. Mater. Rev.* **2003**, *48*, 125–164. [CrossRef]
15. Purtov, J.; Gorb, E.; Steinhart, M.; Gorb, S. Measuring of the hardly measurable: Adhesion properties of anti-adhesive surfaces. *Appl. Phys. A* **2013**, *111*, 183–189. [CrossRef]
16. Pepelyshev, A.; Borodich, F.; Galanov, B.; Gorb, E.; Gorb, S. Adhesion of Soft Materials to Rough Surfaces: Experimental Studies, Statistical Analysis and Modelling. *Coatings* **2018**, *8*, 350. [CrossRef]
17. Galanov, B.A. Models of adhesive contact between rough elastic bodies. *Int. J. Mech. Sci.* **2011**, *53*, 968–977. [CrossRef]
18. Polyakov, B.; Vlassov, S.; Dorogin, L.M.; Kulis, P.; Kink, I.; Lohmus, R. The effect of substrate roughness on the static friction of CuO nanowires. *Surf. Sci.* **2012**, *606*, 1393–1399. [CrossRef]
19. Deng, W.; Kesari, H. Molecular statics study of depth-dependent hysteresis in nano-scale adhesive elastic contacts. *Model. Simul. Mater. Sci. Eng.* **2017**, *25*, 055002. [CrossRef]
20. Medina, S.; Dini, D. A numerical model for the deterministic analysis of adhesive rough contacts down to the nano-scale. *Int. J. Solids Struct.* **2014**, *51*, 2620–2632. [CrossRef]
21. Grierson, D.S.; Flater, E.E.; Carpick, R.W. Accounting for the JKR–DMT transition in adhesion and friction measurements with atomic force microscopy. *J. Adhes. Sci. Technol.* **2005**, *19*, 291–311. [CrossRef]
22. Kendall, K. *Molecular Adhesion and Its Applications*; Kluwer Academic/Plenum Publishers: New York, NY, USA, 2001.

23. Kohn, J.C.; Ebenstein, D.M. Eliminating adhesion errors in nanoindentation of compliant polymers and hydrogels. *J. Mech. Behav. Biomed. Mater.* **2013**, *20*, 316–326. [CrossRef]
24. Gorb, S.; Jiao, Y.; Scherge, M. Ultrastructural architecture and mechanical properties of attachment pads in *Tettigonia viridissima* (*Orthoptera Tettigoniidae*). *J. Comp. Physiol.* **2000**, *186*, 821–831. [CrossRef]
25. Kovalev, A.; Filippov, A.; Gorb, S. Slow viscoelastic response of resilin. *J. Comp. Physiol.* **2018**, *204*, 409–417. [CrossRef]
26. Oliver, W.C.; Pharr, G.M. Improved technique for determining hardness and elastic modulus using load and displacement sensing indentation experiments. *J. Mater. Res.* **1992**, *7*, 1564–1580. [CrossRef]
27. Johnson, K.L. *Contact Mechanics*; Cambridge University Press: Cambridge, UK, 1985.
28. Maugis, D. *Contact, Adhesion and Rupture of Elastic Solids*; Springer-Verlag: Berlin, Germany, 2000.
29. Kalei, G.N. Some results of microhardness test using the depth of impression (in Russian). *Mashinovedenie* **1968**, *4*, 105–107.
30. Khrushchov, M.M.; Berkovich, E.S. *Devices PMT-2 and PMT-3 for Microhardness Testing (in Russian)*; USSR Academy of Sciences Publising: Moscow, Russia, 1950.
31. Borodich, F.M.; Keer, L.M.; Korach, C.S. Analytical study of fundamental nanoindentation test relations for indenters of non-ideal shapes. *Nanotechnology* **2003**, *14*, 803–808. [CrossRef]
32. Kindrachuk, V.M.; Galanov, B.A.; Kartuzov, V.V.; Dub, S.N. On elastic nanoindentation of coated half-spaces by point indenters of non-ideal shapes. *Nanotechnology* **2006**, *17*, 1104–1111. [CrossRef] [PubMed]
33. Bulychev, S.I.; Alekhin, V.P.; Shorshorov, M.K.; Ternovskii, A.P.; Shnyrev, G.D. Determination of Young's modulus according to indentation diagram (in Russian). *Ind. Lab.* **1975**, *41*, 1409–1412.
34. Borodich, F.M. The Hertz-type and adhesive contact problems for depth-sensing indentation. *Adv. App. Mech.* **2014**, *47*, 225–366.
35. ISO. *International Standard ISO 14577-1:2015. Metallic Materials—Instrumented Indentation Test for Hardness and Materials Parameters—Part 1: Test Method*; ISO14577-1; ISO: Geneva, Switzerland, 2015.
36. Giannakopoulos, A.; Suresh, S. Determination of elastoplastic properties by instrumented sharp indentation. *Scr. Mater.* **1999**, *40*, 1191–1198. [CrossRef]
37. Borodich, F.M.; Keer, L.M. Evaluation of elastic modulus of materials by adhesive (no-slip) nanoindentation. *Proc. R. Soc. Lond. A* **2004**, *460*, 507–514. [CrossRef]
38. Chaudhri, M.M.; Lim, Y.Y. Nanoindentation Techniques: A Critical Assessment of the Current Methods of Data Analysis. *Key Eng. Mater.* **2007**, *345–346*, 1107–1114. [CrossRef]
39. Galanov, B.A.; Dub, S.N. Critical comments to the Oliver–Pharr measurement technique of hardness and elastic modulus by instrumented indentations and refinement of its basic relations. *J. Superhard Mater.* **2017**, *39*, 373–389. [CrossRef]
40. Kossovich, E.L.; Borodich, F.M.; Epshtein, S.A.; Galanov, B.A.; Minin, M.G.; Prosina, V.A. Mechanical, structural and scaling properties of coals: depth-sensing indentation studies. *App. Phys. A* **2019**, *125*, 195. [CrossRef]
41. Argatov, I.I.; Borodich, F.M.; Epshtein, S.A.; Kossovich, E.L. Understanding of material properties of thin films attached to substrates: Depth-sensing unloading of elasto-plastic and elasto-brittle materials. *Mech. Mater.* **2017**, *114*, 172–179. [CrossRef]
42. Galanov, B.A.; Grigor'ev, O.N.; Mil'man, Y.V.; Ragozin, I.P. Determination of the hardness and Young's modulus from the depth of penetration of a piramidal indentor. *Strength Mater.* **1983**, *15*, 1624–1628. [CrossRef]
43. Galanov, B.A.; Grigor'ev, O.N.; Mil'man, Y.V.; Ragozin, I.P.; Trefilov, V.I. Determination of the hardness and Young's modulus with elastoplastic penetration of indentors into materials. *Sov. Phys. Dokl.* **1984**, *29*, 146–147.
44. Borodich, F.M.; Keer, L.M. Contact problems and depth-sensing nanoindentation for frictionless and frictional boundary conditions. *Int. J. Solids Struct.* **2004**, *41*, 2479–2499. [CrossRef]
45. Enders, S.; Barbakadse, N.; Gorb, S.; Arzt, E. Exploring Biological Surfaces by Nanoindentation. *J. Mater. Res.* **2004**, *19*, 880 – 887. [CrossRef]
46. Arzt, E.; Enders, S.; Gorb, S. Towards a micromechanical understanding of biological surface devices. *Zeitschrift für Metallkunde* **2002**, *93*, 345–351. [CrossRef]
47. Klein, M.C.; Deuschle, J.; Gorb, S. Material properties of the skin of the Kenyan sand boa *Gongylophis colubrinus* (*Squamata, Boidae*). *J. Comp. Physiol. A* **2010**, *196*, 659–68. [CrossRef]

48. Klein, M.C.; Gorb, S. Epidermis architecture and material properties of the skin of four snake species. *J. R. Soc. Interface* **2012**, pp. 3140–3155. [CrossRef]
49. Huber, G.; Orso, S.; Spolenak, R.; Wegst, U.; Enders, S.; Gorb, S.; Arzt, E. Mechanical properties of a single gecko seta. *Int. J. Mater. Res.* **2008**, *99*, 1113–1118. [CrossRef]
50. Schwendicke, F.; Eggers, K.; Meyer-Lueckel, H.; Dörfer, C.; Kovalev, A.; Gorb, S.; Paris, S. In vitro Induction of Residual Caries Lesions in Dentin: Comparative Mineral Loss and Nano-Hardness Analysis. *Caries Res.* **2015**, *49*, 259–265. [CrossRef] [PubMed]
51. Krings, W.; Kovalev, A.; Glaubrecht, M.; Gorb, S. Differences in the Young modulus and hardness reflect different functions of teeth within the taenioglossan radula of gastropods. *Zoology* **2019**, *137*, 125713. [CrossRef] [PubMed]
52. Rundlöf, M.; Karlsson, M.; Wågberg, L.; Poptoshev, E.; Rutland, M.; Claesson, P. Application of the JKR Method to the Measurement of Adhesion to Langmuir-Blodgett Cellulose Surfaces. *J. Colloid Interface Sci.* **2000**, *230*, 441–447. [CrossRef] [PubMed]
53. Carrillo, F.; Gupta, S.; Balooch, M.; Marshall, S.; Marshall, G.; Pruitt, L.; Puttlitz, C. Nanoindentation of polydimethylsiloxane elastomers: Effect of crosslinking, work of adhesion, and fluid environment on elastic modulus. *J. Mater. Res.* **2005**, *20*, 2820–2830. [CrossRef]
54. Wahl, K.J.; Asif, S.A.S.; Greenwood, J.A.; Johnson, K.L. Oscillating adhesive contacts between micron-scale tips and compliant polymers. *J. Colloid Interface Sci.* **2006**, *296*, 178–188. [CrossRef]
55. Ebenstein, D.M.; Wahl, K.J. A comparison of JKR-based methods to analyze quasi-static and dynamic indentation force curves. *J. Colloid Interface Sci.* **2006**, *298*, 652–662. [CrossRef]
56. Yu, Y.; Sanchez, D.; Lu, N. Work of adhesion/separation between soft elastomers of different mixing ratios. *J. Mater. Res.* **2015**, *30*, 2702–2712. [CrossRef]
57. Goodwyn, P.; Peressadko, A.; Schwarz, H.; Kastner, V.; Gorb, S. Material structure, stiffness, and adhesion: Why attachment pads of the grasshopper (*Tettigonia viridissima*) adhere more strongly than those of the locust (*Locusta migratoria*) (*Insecta: Orthoptera*). *J. Comp. Physiol. A* **2006**, *192*, 1233–1243. [CrossRef]
58. Bußhardt, P.; Wolf, H.; Gorb, S. Adhesive and frictional properties of attachment pads in two species of stick insects (Phasmatoidea) with smooth and nubbly euplantulae. *Zoology* **2012**, *115*, 135–141. [CrossRef]
59. Eichler-Volf, A.; Xue, L.; Kovalev, A.; Gorb, E.; Gorb, S.; Steinhart, M. Nanoporous Monolithic Microsphere Arrays Have Anti-Adhesive Properties Independent of Humidity. *Materials* **2016**, *9*, 1–11. [CrossRef]
60. Xue, L.; Kovalev, A.; Eichler-Volf, A.; Steinhart, M.; Gorb, S. Humidity-enhanced wet adhesion on insect-inspired fibrillar adhesive pads. *Nat. Commun.* **2015**, *6*, 1–9. [CrossRef] [PubMed]
61. Johnson, K.L.; Kendall, K.; Roberts, A.D. Surface energy and the contact of elastic solids. *Proc. R. Soc. Lond. A* **1971**, *324*, 301–313. [CrossRef]
62. Derjaguin, B.V.; Muller, V.M.; Toporov, Y.P. Effect of contact deformations on adhesion of particles. *J. Colloid Interface Sci.* **1975**, *53*, 314–326. [CrossRef]
63. Beach, E.R.; Tormoen, G.W.; Drelich, J.; Han, R. Pull-off force measurements between rough surfaces by atomic force microscopy. *J. Colloid Interface Sci.* **2002**, *247*, 84–99. [CrossRef] [PubMed]
64. Borodich, F.M.; Galanov, B.A. Non-direct estimations of adhesive and elastic properties of materials by depth-sensing indentation. *Proc. R. Soc. London A* **2008**, *464*, 2759–2776. [CrossRef]
65. Sirghi, L.; Rossi, F. Adhesion and elasticity in nanoscale indentation. *App. Phys. Lett.* **2006**, *89*, 243118–243118. [CrossRef]
66. Borodich, F.M.; Galanov, B.A.; Gorb, S.N.; Prostov, M.Y.; Prostov, Y.I.; Suarez-Alvarez, M.M. Evaluation of adhesive and elastic properties of materials by depth-sensing indentation of spheres. *J. App. Phys. A Mater. Sci. Process.* **2012**, *108*, 13–18. [CrossRef]
67. Borodich, F.M.; Galanov, B.A.; Gorb, S.N.; Prostov, M.Y.; Prostov, Y.I.; Suarez-Alvarez, M.M. An inverse problem for adhesive contact and non-direct evaluation of material properties for nanomechanics applications. *Nanoscale Syst. Math. Model. Theory Appl.* **2012**, *1*, 80–92. [CrossRef]
68. Borodich, F.M.; Galanov, B.A.; Gorb, S.N.; Prostov, M.Y.; Prostov, Y.I.; Suarez-Alvarez, M.M. Evaluation of Adhesive and Elastic Properties of Polymers by the BG Method. *Macromol. React. Eng.* **2013**, *7*, 555–563. [CrossRef]
69. Boyd, S.; Vandenberghe, L. *Convex Optimization*; Cambridge Univ. Press: Cambridge, UK, 2004.
70. Chong, E.; Zak, S. *An Introduction to Optimization*; Wiley: New York, NY, USA, 2001.
71. Tabor, D. Surface forces and surface interactions. *J. Colloid Interface Sci.* **1977**, *58*, 2–13. [CrossRef]

72. Muller, V.M.; Yushchenko, V.S.; Derjaguin, B.V. On the influence of molecular forces on the deformation of an elastic sphere and its sticking to a rigid plane. *J. Colloid Interface Sci.* **1980**, *77*, 91–101. [CrossRef]
73. Domke, J.; Radmacher, M. Measuring the elastic properties of thin polymer films with the atomic microscope. *Langmuir* **1998**, *14*, 3320–3325. [CrossRef]
74. Dimitriadis, E.K.; Horkay, F.; Maresca, J.; Kachar, B.; Chadwick, R.S. Determination of elastic moduli of thin layers of soft material using the atomic force microscope. *Biophys. J.* **2002**, *82*, 2798–2810. [CrossRef]
75. Sirghi, L.; Ponti, J.; Broggi, F.; Rossi, F. Probing elasticity and adhesion of live cells by atomic force microscopy indentation. *Eur. Biophys. J.* **2008**, *37*, 935–945. [CrossRef] [PubMed]
76. Zhu, X.; Siamantouras, E.; Liu, K.K.; Liu, X. Determination of work of adhesion of biological cell under AFM bead indentation. *J. Mech. Behav. Biomed. Mater.* **2016**, *56*, 77–86. [CrossRef] [PubMed]
77. Bouchonville, N.; Meyer, M.; Gaude, C.; Gay, E.; Ratel, D.; Nicolas, A. AFM mapping of the elastic properties of brain tissue reveals kPa μm(-1) gradients of rigidity. *Soft Matter* **2016**, *12*, 6232–6239. [CrossRef]
78. Notbohm, J.; Poon, B.; Ravichandran, G. Analysis of nanoindentation of soft materials with an atomic force microscope. *J. Mater. Res.* **2012**, *27*, 229–237. [CrossRef]
79. Lin, D.C.; Dimitriadis, E.K.; Horkay, F. Robust strategies for automated AFM force curve analysis-I. Non-adhesive indentation of soft, inhomogeneous materials. *J. Biomech. Eng.* **2007**, *129*, 430–440. [CrossRef]
80. Lin, D.C.; Dimitriadis, E.K.; Horkay, F. Robust strategies for automated AFM force curve analysis-II. Adhesion-influenced indentation of soft, elastic materials. *J. Biomech. Eng.* **2007**, *129*, 904–912. [CrossRef]
81. Bouchonville, N.; Nicolas, A. Quantification of the Elastic Properties of Soft and Sticky Materials Using AFM. *Methods Mol. Biol.* **2019**, *1886*, 281–290.
82. Perepelkin, N.; Argatov, I.; Borodich, F. Evaluation of elastic and adhesive properties of solids by depth-sensing indentation (in press). *J. Adhes.* **2019**, doi:10.1080/00218464.2019.1686981. [CrossRef]
83. Argatov, I.I.; Sabina, F.J. Asymptotic analysis of the substrate effect for an arbitrary indenter. *Q. J. Mech. Appl. Math.* **2013**, *66*, 75–95. [CrossRef]
84. Argatov, I.I.; Borodich, F.M.; Popov, V.L. JKR adhesive contact for a transversely isotropic layer of finite thickness. *J. Phys. D Appl. Phys.* **2016**, *49*, 045307. [CrossRef]
85. Borodich, F.M.; Galanov, B.A.; Perepelkin, N.V.; Prikazchikov, D.A. Adhesive contact problems for a thin elastic layer: Asymptotic analysis and the JKR theory. *Math. Mech. Solids* **2019**, *24*, 1405–1424. [CrossRef]
86. Erbaş, B.; Aydın, Y.E.; Borodich, F.M. Indentation of thin elastic films glued to rigid substrate: Asymptotic solutions and effects of adhesion. *Thin Solid Film.* **2019**, *683*, 135–143. [CrossRef]
87. Sergici, A.O.; Adams, G.G.; Müftü, S. Adhesion in the contact of a spherical indenter with a layered elastic half-space. *J. Mech. Phys. Sol.* **2006**, *54*, 1843–1861. [CrossRef]
88. Stan, G.; Adams, G.G. Adhesive contact between a rigid spherical indenter and an elastic multi-layer coated substrate. *Int. J. Sol. Struct.* **2016**, *87*, 1–10. [CrossRef]
89. Jin, F.; Guo, X.; Gao, H. Adhesive contact on power-law graded elastic solids: The JKR–DMT transition using a double-Hertz model. *J. Mech. Phys. Sol.* **2013**, *61*, 2473–2492. [CrossRef]
90. Sridhar, I.; Johnson, K.L.; Fleck, N.A. Adhesion mechanics of the surface force apparatus. *J. Phys. D Appl. Phys.* **1997**, *30*, 1710–1719. [CrossRef]
91. Johnson, K.L.; Sridhar, I. Adhesion between a spherical indenter and an elastic solid with a compliant elastic coating. *J. Phys. D Appl. Phys.* **2001**, *34*, 683–689. [CrossRef]
92. Sridhar, I.; Zheng, Z.W.; Johnson, K.L. A detailed analysis of adhesion mechanics between a compliant elastic coating and a spherical probe. *J. Phys. D Appl. Phys.* **2004**, *37*, 2886–2895. [CrossRef]
93. Perepelkin, N.; Kovalev, A.; Gorb, S.; Borodich, F. Estimation of the elastic modulus and the work of adhesion of soft materials using the extended Borodich-Galanov (BG) method and depth sensing indentation. *Mech. Mat.* **2019**, *129*, 198–213. [CrossRef]
94. Ahn, S.J. *Least Squares Orthogonal Distance Fitting of Curves and Surfaces in Space*; Springer: Berlin, Germany, 2004.
95. Boggs, P.T.; Byrd, R.H.; Schnabel, R.B. A stable and efficient algorithm for nonlinear orthogonal distance regression. *SIAM J. Sci. Stat. Comput.* **1987**, *8*, 1052–1078. [CrossRef]

MDPI
St. Alban-Anlage 66
4052 Basel
Switzerland
Tel. +41 61 683 77 34
Fax +41 61 302 89 18
www.mdpi.com

Nanomaterials Editorial Office
E-mail: nanomaterials@mdpi.com
www.mdpi.com/journal/nanomaterials

www.ingramcontent.com/pod-product-compliance
Lightning Source LLC
Chambersburg PA
CBHW062056220926
51850CB00010B/2792
* 9 7 8 3 0 3 9 4 3 0 9 2 5 *